EDWARD F. BENSON

THE CHALLONERS

Elibron Classics
www.elibron.com

Elibron Classics series.

ISBN 1-4021-6774-1 (paperback)
ISBN 1-4212-8237-2 (hardcover)

This Elibron Classics Replica Edition is an unabridged facsimile of the edition published in 1904 by Bernhard Tauchnitz, Leipzig.

EACH VOLUME SOLD SEPARATELY.

COLLECTION OF BRITISH AUTHORS

TAUCHNITZ EDITION.

VOL. 3752.

THE CHALLONERS BY E. F. BENSON.

IN ONE VOLUME.

LEIPZIG: BERNHARD TAUCHNITZ.

PARIS: LIBRAIRIE H. GAULON & Cie, 39, RUE MADAME;
THE GALIGNANI LIBRARY, 224, RUE DE RIVOLI.

(See also pp. 3–6 of Large Catalogue.)

COLLECTION

OF

BRITISH AUTHORS

TAUCHNITZ EDITION.

VOL. 3752.

THE CHALLONERS. By E. F. BENSON.

IN ONE VOLUME.

TAUCHNITZ EDITION.

By the same Author,

DODO	1 vol.
THE RUBICON	1 vol.
SCARLET AND HYSSOP . .	1 vol.
THE BOOK OF MONTHS . .	1 vol.
THE RELENTLESS CITY . .	1 vol.
MAMMON & CO.	2 vols.
AN ACT IN A BACKWATER .	1 vol.
THE IMAGE IN THE SAND .	2 vols.
THE ANGEL OF PAIN . . .	2 vols.

THE CHALLONERS

BY

E. F. BENSON

AUTHOR OF "DODO," "THE RELENTLESS CITY,"
"MAMMON & CO.," ETC.

COPYRIGHT EDITION

LEIPZIG

BERNHARD TAUCHNITZ

1904.

"O World as God has made it! All is beauty:
And knowing this, is love, and love is duty.
What further may be sought for or declared?"

"*The Guardian Angel*,"—R. BROWNING.

THE CHALLONERS.

CHAPTER I.

THE hot stress of a real midsummer day towards the end of June had given place to the exquisite tempered warmth of evening, and a little breeze born of the hour before sunset, and made fragrant among the glowing flower-beds of the vicarage garden, just ruffled the hair of Helen Challoner as she half sat, half lay in a long deck-chair at the edge of the croquet-lawn, reading a red-covered book with the absorbed intentness which she devoted to any occupation that interested her. To the west a line of tall box-hedge, of that smooth and compacted growth which many years alone can give, screened her from the level rays of the sun, which was but an hour above the horizon, and performed the almost more desirable function of screening her from the windows of the house, for a cigarette was between her fingers, and the juxtaposition of women and tobacco was a combination that had probably never occurred to her father as possible. The cigarette, however, was, as a matter of fact, wasting its sweetness uninhaled, and burning down with a long peninsula on the leeward side of it of charred paper, for her book absorbed her quite completely. Indeed, this seat here under cover of the box-hedge was a manœuvre of double strategy, for the book was no less anathema in this house than the cigarette, being, in fact, "The Mill on the Floss," by an author who, however celebrated, yet remained, in the opinion both of Helen's father and aunt, a person of unchristian belief and heathenish conduct.

Helen wore no hat, and the dusky smouldering gold of her hair burned low over her forehead. Her eyelids, smooth with the unwrinkled firmness of flesh of twenty-two years, drooped low over her book, but between the lids there showed a thin line of matchless violet. There were but a few more pages to read; and her underlip, full and sensitive in outline, quivered from time to time with the emotion that so filled her, and her breath came quickly through her thin nostrils. As she read on, her half-smoked cigarette dropped from between the fingers of her left hand, and sent up little whorls of blue smoke as it lay unheeded on the grass, and her eyes grew suddenly dim. Then the last page was turned, and with a sudden sobbing intake of her breath she closed the book.

She sat quite still for a moment, the book lying in her lap, looking with misty unseeing eyes over the great stretch of open land and sky in front of her. In the immediate foreground lay the croquet lawn, with disjected mallets and aimless balls scattered about, while slowly across it, like some silent tide, the shadows grew and lengthened. Beyond, at the top of a grassy bank still in sunlight, ran a terraced walk, bordered deeply with tall herbaceous plants; further, out of sight behind the border, were a few fields, water meadows of the chalk-stream, and beyond again and above rose the splendid and austere line of Hampshire downs, tanned with this month of English summer to a russet mellowness. A sky of untarnished blue held a slip of pale and crescent moon, and the splendour and the unutterable sadness of evening, of a day gone, brooded, a sweet regretful presence, over everything.

Suddenly the girl sat up.

"Martin!" she cried, "Martin!"

"Well?" asked a very lazy voice from a hammock between two trees at the end of the lawn.

"Come here. Oh, do come. I can't shout."

The hammock ropes wheezed and creaked, and a tall, loose-limbed boy, looking not much more than twenty, strolled over to where she sat.

"I've won my bet," he said. "So pay up, Helen. I said the end would make you cry. You are crying, you know. I count that crying."

"I know. I'll pay all right," she said. "I almost wish it had been more."

"So do I," said Martin. "That's easily arranged then."

Helen paid no attention to this.

"Oh, Martin, those two coming together like that at the end. And that beast, that beast——"

"Stephen?"

"Yes, among others. But Tom particularly. They none of them knew, they none of them guessed what she, what Maggie was. Oh, oh! How horribly sad, and how horribly beautiful—like, like this evening."

Martin took out his cigarette case.

"For you?" he asked.

"No, you gave me one which I haven't—I don't know where it is. Oh, it's smoking itself on the grass. Oh, my goodness! Anyhow, Maggie lived; that is the point. Dreadful people, dreadful circumstances, all that one would think would make living impossible, surrounded her. But she managed it. And what am I to do, please?"

Martin laughed.

"I wonder if you know how like you that is?" he said.

"What is?"

"Your instant application of Maggie to yourself. Really, it is very odd that you and I are twins. If only I had half your eye for the practical way of getting through things I should pass my examinations. And if you had had only half my eye for the theoretical beauty of leaving distasteful things alone——"

Helen sat up with a quick decisive movement, letting the book drop on the grass.

"Martin, if we didn't happen to have been brother and sister we should have fallen desperately in love with each other and been accepted at once. At least, I should have proposed to you, and you would certainly have said 'Yes.' And I should have made home happy for you on two-pence farthing a year, and always had your slippers warm when you came home in the evening, and the kettle boiling on the hob. And you could have spent the rest of our joint incomes on grand pianos and music paper."

"You are too overwhelmingly generous, Helen," said he. "I don't think I can accept it from you."

Helen got up.

"Oh, how I hate, how I hate——" she began.

"That's no use," said Martin.

"Use? Of course not. Oh, it's all very well for you; you are away half the year at Cambridge, and have no end of a time. But I am here. I and the Room!"

"What's the 'Room?'" asked her brother.

Helen pushed back her hair again and sat down on the lawn by Martin.

"The Room is the latest of my many trials," she said. "It is quite new. Outside it is corrugated iron, inside it is distemper, covered by a dreadful sort of moisture, which is essence of village children. On the walls there are maps of the Holy Land and Hampshire. I know the road from Dan even unto Beersheba as well as I know the road from here to Winchester. There is a library there of soiled books of travel and missionary enterprise, and a complete set of 'Good Words.' There is also a well-spring there, only I can't find it and stop it up, which continually pours up an odour of stuffiness. It is the sort of place where nothing nice could ever happen. And there on Tuesday evening I teach arithmetic to dreadful little boys. On Wednesday I read to mothers—I am getting to hate the word—who knit shapeless articles while I read. I read them abominable little stories about the respective powers of faith, hope and love, and the virtue of being good, and the vice of being wicked. I don't suppose any of them could be wicked if they tried."

Helen paused a moment.

"Oh, Martin, it is heavenly to have you at home, and be able to say all these things straight out just once. It makes me feel so much better. May I go on?"

"Yes; take your time," said Martin.

"Well, where had I got to? Oh yes, Wednesday. On Thursday Mr. Wilkins—he's the new curate, whom you haven't seen yet: spectacles, bicycle, and proposes to me every now and then—Mr. Wilkins on Thursday has something for men only; I don't know what, but I'm sure it's dreadful. Friday—girls' class. And on Saturday a choir practice. A—choir—practice. Now you have been to church here——"

"Rather," said Martin.

"——and heard the singing. It is to produce that marvellous result that we practice. Even I know how awful it is. There was a man called the Reverend P. Henley. I sing the alto of his horrid chant. Would you like to hear me sing? And on Sunday I have the Sunday school. They use heaps of pomatum, you know. And they learn by heart their duty towards their neighbours, and when I am not looking pull each other's hair. Then it is Monday again, and we begin all over again. Oh, think of it! You see I am not by nature a ministering angel, and I have to spend my whole life in ministering to these people. They have no intelligence, nothing that I can lay hold of or join hands with. It is not their fault, and it is not my fault that I am not a ministering angel. But what is the use of battering at their intelligences when they haven't got any? Also, they are personally distasteful to me."

Martin laughed at this tirade, and thoughtfully executed a gnat that had designed to dine off his brown fingers.

"Why, I thought you were such a success," he said. "Father held you up to me as an example and a shining light."

"Of course I'm a success," cried Helen. "I've got to do this sort of thing, and if one has to do something it is simple imbecility not to do it well. You're an imbecile, you know, darling."

"Oh, I know that," said he. "At least I've been told it often enough."

Helen was silent a moment, looking very affectionately

at her brother's long slim figure as he lay stretched on the grass by her side. His straw hat was tilted over his eyes, and of his face there appeared only his chin and his mouth a little open, showing a very white line of teeth. And the current of her thoughts hardly changed when she went on to speak of him, not herself.

"Martin, how is it you can't get through your examinations?" she asked. "You do work, don't you? And though I called you an imbecile just now, you have more perception than most people. Or do you spend all your day at the piano?"

"He has forbidden me to have a piano in my rooms next term," said Martin. "So I shall have to waste more time in walking to the pianos of other people, and interrupting their work as well as my own."

"Ah, that's too bad," said Helen.

Martin only grunted in reply, and his sister went on:

"But it is foolish of you," she said. "Indeed it is foolish. No doubt what you have got to do, Greek, Latin, is all very dull to you and seems very useless, but it is surely better to look at it as one of those things that has got to be done. As you say, and as father says, and as I say, I am a success at all these dreadful functions in the Room. Why? Merely because it has got to be done, and therefore, although it is all intensely stupid and bores me so much that I could cry, I attend sufficiently to do it respectably. Now, can't you adopt the same attitude towards classics? Besides, you know what father feels about it."

"I am perfectly aware of what father feels about it," said Martin drily.

"Has he been at you again?"

"Yes, I think you might call it that without conveying

a false impression. He apparently wants to give me to understand that it is some moral crime not to be able to do Greek Iambics. Well, I am a criminal then. I can't. Also, that it is impossible to be educated without. Then I began arguing, which is always stupid, and said I supposed it depended on what one meant by education. And he said he imagined he was the best judge of that. So there we were!"

"And what do you mean by education?" asked Helen.

"Why, of course, the appreciation of beauty," said Martin quickly. "'O world as God has made it'—you know the lines."

"Ah, say them," she said.

Martin sat up, tilting back his hat.

> "'O world as God has made it, all is beauty;
> And knowing this is love, and love is duty.
> What further may be sought for or declared?'"

"Yes, that isn't a bad creed," said Helen.

"I hope not, for it is mine. And it seems to me that you may look for beauty and find it in almost everything. Where you look for it should depend entirely on your tastes. Father finds it in the works of Demosthenes, but I in the works of Schumann and a few other people he has never heard of."

"But aren't Greek plays beautiful?" asked Helen.

"Oh, I daresay. But being what I am, music concerns me more. Don't let's argue. It is so enfeebling. When I begin arguing I always feel like Mr. Tulliver when he said, 'It's puzzling work, is talking!'"

Helen laughed.

"Well, you and I ought to be pretty well puzzled by

now," she said. "I'm sure we've talked enough. I'll play you one half game of croquet before dinner. Oh, by the way, father is dining with Uncle Rupert. You and Aunt Clara and I will be alone. You will have to read prayers."

"And sing the hymn an octave below," remarked Martin.

The Honourable and Reverend Sidney Challoner—or as he preferred to be addressed, the Reverend and Honourable—was a man of method and economy, who hated wasting anything, from time down to the brown paper in which parcels arrived; and at this moment he was employing the half-hour before it was necessary to go to dress for dinner at Chartries, his brother's place, which stood pleasantly among woods about a mile distant, in finishing his sermon for Trinity Sunday. His study where he worked was singularly like himself, and seemed as integral a part of him as the snail-shell is of the snail. There was nothing, for instance, in the least drowsy or dusty about the room. Everything was in its place, the place of each thing being in every case strictly determined by the use to which it was to be put, and the frequency with which it was to be used. A scrupulously orderly and energetic severity, in fact, was the keynote of the room.

Something of the same characteristics also ran through the sermon at which he was working, which was an exposition, historically introduced, of the less encouraging and comfortable verses of the Athanasian Creed, which his congregation would have recited during the service. He was master of a style of English in itself neat, correct and lucid, which served him, not as in the sermons of so many preachers, to clothe and cover his lack of ideas, but to reveal the abundance of them, and convey without possibility of misunderstanding but rather with the precision of hitting a nail on the head, what he thought on any particular subject. There survived in him indeed a full if not a double portion of the Puritan spirit on religious matters; and though his mind, his soul, his actions were all dictated and impelled by a fervent and whole-hearted Christianity, yet his eloquence was wont to dwell, and did so here, on the doctrine of eternal damnation with a very curious gusto. It appeared to him that the truth of it was abundantly warranted in the Bible, and that it was therefore his duty as minister of the Word to bring this as well as other doctrines home to his flock. And something of the same grim aspect of duty extended to affairs of his secular life; and where censure was clearly deserved, any offence was visited by him with a force that his approbation sometimes lacked, if there was nothing to blame. The Puritan too survived in a certain mistrust he had of mirth and gaiety: without being in the least sour he was so intensely serious that at any given moment it appeared to him that there was probably something better to do than to laugh, and a moment's thought easily discovered what it was. Of work he was insatiable: if he was unsparing to others at any rate he never spared himself, and the day of rest was to all in his house the most iron day of all. All pleasure, except that which was to him the greatest pleasure in life, active religious work and religious exercises, was put away; but since all exercises, even religious ones, are fatiguing, it was a weary household that went up to bed on Sunday evening.

Now, though to have a very strong vocation towards a particular work, to be convinced that such work is the highest and best in the world, and to do it, is a disposition of affairs that makes for happiness, it is probable that if you had taken Mr. Challoner unawares and asked him if he was happy, he would have hesitated before he answered. For in spite of his firm and convinced attitude both towards life in general and to those most intimate with him, there rose deep down in the man a great fountain of tenderness, a great longing for love. Herein lay the secret tragedy of his life; he longed with the same intensity with which he served God for the ordinary human affections and relationships, but through the armour-crust of his nature, an armour, be it noted, of welded and hammered work and duty, his human hand could not break its way to clasp the hands of others. That still was the tragedy of his life with regard to his two children just as it had been even more bitterly so with regard to his wife, a half Italian by birth, whom he had adored with that serious fervour which suffused his nature. It was just his spiritual anxiety and care for her which had, by a refined irony of fate, come like an impassable barrier between them. To her he seemed always to be checking the innocent and sunny impulses of joy that were as vitally hers as fervour was his. He put it that there was always something better to be done with the precious passing hours than to sing or laugh or gather flowers or embroider some dainty fragment of personal embellishment. Or rather let her take these innocent tastes and raise them, elevate them, dedicate them. Let her sing by all means, but let her gift of music be devoted to the help of the parish choir; let her gather flowers to send to the sick; let her embroider an altar-cloth. But poor Mrs. Challoner, a girl still in years, whose motor power in life was joy, found that to fit her pleasures to useful ends meant that they ceased to be pleasures. There are many natures, not necessarily shallow or selfish, like that; and when her husband told her that the flowers with which she loved to fill her rooms were beautiful to her, so that thereby her thoughts might be led heavenwards, she was minded to throw them away.

From the first indeed the marriage had been strangely ill-assorted; it may have been made in heaven, but in that case it would probably have been far better if it had not come down to earth. Sidney Challoner had had his reason and his senses taken captive for a time by this delicious piece of dew and sunlight; on her side his imperiousness, his eager, over-mastering desire for her, his extreme good looks, and perhaps also the fact that he stood next in succession to the Earldom of Flintshire, his elder brother, the present holder of that delightful position, being unmarried, led her to accept his devotion. Then disillusionment had soon come to each; the exquisite child-like beauty of his wife, behind which he had conjectured the child-like spirit, he found to be a mere mask; while to her the fiery dominating lover turned to a hard unbending master. A year after their marriage twins were born, and from that time the girl-mother had drooped and dwindled. The fogs of this northern climate, fogs too more intimate and distressing of mind and spirit, the absence of mirth and laughter, chilled her to the bone, and a year afterwards she was dead.

Her death left him inconsolable in so far that he determined never to marry again; but when his sister Clara came to keep house for him and look after the early education of Martin and Helen, it cannot be denied that the widower found himself more comfortable than he had been. For Clara was one of those not uncommon English spinsters who had a perfect passion for doing the things she ought to do, and leaving completely undone the things she ought not. As the feminine element in the house of a parish priest it was her clear mission to be aunt if not mother to the flock; and classes and instructions, so hated of her niece's soul, grew up under her care like seed sown in April. She had practically no pleasures, and her only relaxation was Patience, which she played regularly from the time dinner was finished till family prayers at a quarter to ten. Precisely at twenty minutes to ten, if the cards were going awkwardly, she began to cheat, and continued, if necessary, to cheat, until the parlour-maid began to set out a row of chairs for the servants. Thus she was able by the time they filed in to sweep the cards triumphantly up together in their due and proper order, and be humbly thankful for the temptations into which she had not fallen that day.

Mr. Challoner this evening found that the peroration with which he concluded his sermon took rather less time than he had anticipated, and there was still some ten minutes after he had arranged the sheets in order and placed them under a paper weight to be read through in the morning, before he need go to dress. As his custom was he closed his eyes for a moment after finishing his work, in silent prayer that it might bear good fruit, and then, hearing the clash of croquet balls from the garden, he strolled out to see his children. He had had a very unpleasant talk with Martin that morning on the subject of his late failure at Cambridge; and though the occasion seemed to him then, and seemed still, to have demanded stern speaking, he had wondered several times since whether he had not been too severe. Yet how else, except by very earnest remonstrance, could he awaken in the lad his sense of responsibility with regard to the spending of the days that would never come again? All his life he had faithfully and strenuously striven to implant in his boy the duty of making the best and the most of his youth. Prayer and work were the two great guides of life: these must be constant and concentrated, and how gravely and mortally would he himself be to blame if through any want of inculcation on his part his son grew up tepid in the one, and slack in the other! Still, and here his essential tenderness groped about, Martin was young yet, and more tender perhaps in mind even than in years, and the clash of croquet balls and a sudden burst of boyish laughter from the lawn, made him long to enter into his children's pleasures. So without putting on his hat, for the evening breeze was not too cool to the head, he went out down the box-hedge and round the corner on to the croquet lawn.

Martin, standing with his back to him, had not heard his approach, and was examining the position of his two balls, which were quite close together but with an uncompromising wire between them. On the bank where they had been sitting lay "The Mill on the Floss," and Helen was standing close by her brother in the proud calm consciousness of having wired him with complete success.

"Well, of all the devilish things to do, Helen," said Martin at length, and struck wildly in the hopes of an impossible cannon off the wire.

"My turn, I think," she said.

She walked across to the ball in play, and saw her father.

"Come and play, father?" she said.

"No, dear, thanks. I must go and dress in a few minutes. Martin, old boy, come here a moment."

Again his duty, the need for remonstrance strove with his tenderness.

"Martin," he said gently, "that's rather strong language to use to your sister, isn't it? Don't get into that sort of habit, dear fellow: never use words idly like that."

At this all the genial instinctive pleasure faded out of Martin's face and his eyes fell.

"Yes, father, I'm sorry," he said, in a perfectly dull conventional voice.

"I know it was only thoughtlessness, old boy," said his father, "but try to think. There then. How's the game going?—is Helen playing with the frightful precision we are getting accustomed to? Look there, she's hit your ball from right across the lawn. Don't be too merciless, Helen, with your poor brother."

Helen smiled, and made some laughing reply to her father. Then her eye caught sight of the book lying on the bank, her smile faded, and as she went after the ball she had hit she wondered what could be done. She guessed, though she had not heard the words, that Martin had already been rebuked for what he had said: she knew there had been one dreadful hour already that morning, and another was certain if her father saw the book. Meantime he was strolling down the lawn right in its direction, where it was lying radiant and blatant in its crimson cover on the vivid green of the grass. Martin also had seen what would happen, and as she passed him whispered to her:

"He'll see it. O Lord!" with a drearily comic expression.

Mr. Challoner strolled on, came to where the book lay, and picked it up with the amiable intention of putting it on the chair to save its cover from the damp. As he did this, he read the title on the back. Then there was a dreadful pause.

"Is this yours, Martin?" he asked.

"Yes, father."

Mr. Challoner said nothing more, but went on his way, taking the book with him. At the corner of the box-hedge, however, he turned.

"If you are up when I come back, Martin," he said, "will you come into my study? But don't wait up for me if I am late."

He turned his back again to walk on, and Martin thought he had gone. But next moment he paused again, and raised his voice slightly.

"You should answer when I speak to you," he said.

"I thought you had gone, sir," said Martin, with a little tremor of irritation in his tone.

This time he passed out of sight, and Martin threw down his croquet-mallet.

"Rather bad luck," he said. "I'm not popular to-day. Helen, what a fool you were to leave it on the grass!"

"Oh, I am so sorry, Martin," she said. "What can I do? Would it do any good if I said I had been reading it?"

"No, not the slightest," said he. "There would be enough to go round."

"I will if you like," said she. "You see the worst of it is that only three days ago, the day before you came home, he said that he would not have a book of hers in the house. But you couldn't be expected to know that."

"No, but I did," said Martin, "because you told me."

Helen threw down her mallet too.

"Oh, it's dreary," she said.

Lord Flintshire, Mr. Challoner's elder brother, with whom he was dining to-night, was a figure of some distinction. He had been at one time a political factor of great weight in the country, a weight due chiefly to the force of inertia, since he never professed the least personal interest in politics, and could not possibly be considered as having any ambition or aim to gratify in spending so much time and labour in the interests of the Conservative party. His wealth and position, in fact, were like a large heavy parcel strongly tied up and dropped into the Tory scale. But at the age of fifty-five he considered that he had done enough, resigned the Cabinet appointment he held, and for the last seven years had devoted himself with far more zest than he had ever brought into the political arena to the aristocratic pursuit of doing nothing whatever. To the successful discharge of this he brought all his acuteness and perception, and practised it with such charming success as to raise it to the level of a fine art. He was never in a hurry, and never either felt or exhibited the slightest sign of irritation or annoyance at anything which the world or the powers of heaven or hell chose to do. He had great appreciation of the fine arts, and even a higher appreciation of the inimitable comedy of life; so that to live in a beautiful house, which he did, and fill it with congenial people, constituted for him a far more engrossing occupation than politics had ever been. For his brother Sidney he had a very real affection, but also a certain sympathetic pity. He could understand, as he had once told him, what it must be to "feel like that."

"You live perpetually in a bracing climate, my dear fellow," he had said, "and find it positively necessary to do moral dumb-bells all day. Yes, I will certainly give you a hundred pounds for your village room. I shall be charmed to do so, but I don't want to hear about it. And pray let me know if you want more."

There was only a small party that night, and when the women went upstairs and the men seceded to the smoking-room, Lord Flintshire detained his brother for a moment as he was leaving.

"Will you not stop a quarter of a hour, my dear fellow," he said, "and have a chat? I have not seen you since Easter. How are you all? How are Helen and Martin? That girl grows handsomer every time I see her. And Martin?"

"Martin has just achieved one of his annual failures at Cambridge," said his father. "Yes, I will wait a quarter of an hour, Rupert. I should like to talk to you about him. I am a good deal troubled."

"Wild oats of some kind?" asked the other. "If so, I should, if I were you, look very steadily in another direction. As one grows older, my dear Sidney, one is apt to look on wild oats as something much more poisonous than they really are—nightshade—deadly nightshade, for instance. But they are only wild oats really."

Martin's father sat down.

"Ah, you don't expect me to share that view," he said. "Sin is sin whether you are twenty or sixty. But Martin, as far as I know, has not been——"

"Playing about," said Lord Flintshire, with the amiable desire to find a periphrasis. But it did not please his brother.

"I can't discuss things with you in that spirit," he said. "However, that point is really alien. I have no reason to suspect Martin of such things. But what I deplore is his general slackness. It is to the mind like low physical health to the body; it predisposes to all diseases. I had to speak to him severely about his failure at Cambridge this morning, too severely perhaps, and this evening again he has distressed me very much."

"What has he done?" asked Rupert.

"Well, you will think it very insignificant, no doubt, but to me it appears most significant of his general state. He was playing croquet with Helen, and I heard him say to her, 'Well, of all the devilish things to do!' Now when we were boys, Rupert, we didn't say that sort of thing at all, and we couldn't have said it to our sisters."

Lord Flintshire felt some kindly amusement at this. Sidney was such a dear fellow.

"But it is some years since we were boys," said he at length, "and rightly or wrongly the world has begun to take things more—how shall I say it?—to ride life on the snaffle instead of the curb. What else has Martin done?"

"He has brought into the house 'The Mill on the Floss.'"

Rupert's admirable courtesy enabled him not to smile.

"Have you read the book?" he asked.

"No, but I will not have a book of that author in the house. I said so only the other day. Martin must have known it. For all I know, he has given it to his sister to read."

"I hope so," said Lord Flintshire quietly. "Because it is a very beautiful book. Of course his disobedience to your wishes is a different point, and to my mind a more serious one. But am I to understand that you are consulting me as to what general line you should take with Martin, what policy you should pursue?"

"Yes, I am very much puzzled, and I cannot seem to

get any guidance about it. It does no good, I am afraid, to pull the poor lad up first here and then there thirty times a day. And it appears to do no good either to talk to him on the general principles of earnestness and industry. But I do so want him to grasp them. All the faults I see in him spring from slackness. He will not think—he did not think what the word he used to his sister means. He never thinks how just a little carelessness about his work repeated and again repeated must lead to a habit of idleness. I am most deeply thankful that our father was strict with us, Rupert. He made industry a habit with one."

Rupert laughed.

"A habit from which I have succeeded in freeing myself," he said. "But Martin is not slack about everything. He is not slack about music."

"Ah, that is a distraction which is responsible for a great deal of his idleness," said his father. "But I have forbidden him to have a piano in his room next term."

Lord Flintshire did not pursue this; there was a plot already on foot here, and his brother got up and with his quick neat touch put straight a couple of books lying on the table.

"There is this too," he said; "not only does my continual correction of him seem to do no real good, but it certainly does harm to my relations with the boy. He will get to look on me as a continual menace to his pleasure, as a continual schoolmaster. And I want to be kind to the lad, to make him happy, to make a friend of him. But when that which I consider my duty leads me to correct him, and again and again to correct him, I am so afraid that his estimate of the love I bear him will be lowered, eclipsed. And nothing in the world, Rupert, could be sadder to me than that my children should not think of me as their friend."

His strong tender voice quivered for a moment as he spoke these words, and he paused to regain the complete control of himself.

"But nothing, not even that," he said, "must or shall stand in my way or count for anything in regard to the responsibility which God has laid upon me to make my children worthy children of Him. I should be the weakest and most culpable of fathers, if for the sake of any human affection, however sweet, I sacrified one jot or tittle of that."

Rupert was silent a moment. Though he had always felt great respect and esteem and strong affection for his brother, he had never found him, emotionally speaking, particularly interesting. He had the greatest admiration for his industrious, strenuous life, his undoubted mental gifts, his swift and keen intelligence, the absolute undeviating probity of his character; but his admiration had been somewhat of the sort a mechanician may feel for his bright engine with its rhythmical accuracy, its precise strokes, its clean efficiency and strength. But suddenly the engine had developed a human and pathetic side; its throbs were not steam-driven only, but they were the throbs of a human heart. True, he had known the wild adoration of Sidney for his girl-wife; but that, with its speedy disillusionment, had seemed to him the one concession Sidney had made to the flesh. It was human, but it was not high humanity, otherwise he would have made a better recovery, so to speak. His passion had been awakened then, but not the man, and his religion and his passion had mixed no better together than oil and water. The experience had not humanised him.

Lord Flintshire's strong appreciation of the inimitable comedy of life did not help him here, as he sat silent for a moment before replying. Elements of comedy were not wanting, his brother's heartfelt distress at the fact of Martin calling his sister "devilish," for instance, was ludicrous enough, but these things combined to form nothing to laugh at; the result was tragedy, tragedy in no grand and great style, but a pitiful little tragedy of misunderstanding and estrangement. And Rupert, knowing his brother and knowing Martin, saw no possibility of comedy entering with any unexpected "happy ending." For Sidney was, so to speak, an irreconcilable; he admitted no sort or shadow of compromise; he would hold no parleying with the enemy, even if the enemy was entrenched in one of his own household. He and Martin in fact disagreed vitally and fundamentally; the lad was a good lad accidentally; essentially, he was an artist to his finger tips. Those were the influences which governed him. But to his father all the artists and all the artistic achievements of man were no more than a fringe on the visible garment of God.

"No one can really help you in this," said Rupert at length, "except yourself and Martin. But I can suggest to you a certain point of view. Do, I beg you, allow for individualism in other people. You yourself, dear Sidney, have a great deal of it. But there is no reason to suppose that Martin has any less. And remember also that the younger generation is always ahead of the elder, and though we can, by using extreme care, influence them a little, yet the reins of government are in their hands, not ours. That is partly why I retired from politics. And as a practical suggestion I offer you this—I beg you to say nothing more about the 'Mill on the Floss' to Martin. It is quite impossible that he should agree with you, simply because he is of the next generation to you. Indeed, if you do not take care, that which you are afraid of will certainly happen, even if it has not happened already. He will get to think of you as a man who is always finding fault, always correcting—a thing fatal to friendship."

"Is it irremediable if it has already happened?" asked Sidney, with a rather pathetic humility.

"Of course it is not, just because boys are so extraordinarily generous, so eager to like one. Martin is a delightful boy: he is upright, honest, clean. Be thankful for that, and let him develop on his own lines. He will do so, by the way, whether you like it or not, so it is just as well to like it. Besides, you must not interfere with other people's individualities. I feel that rather strongly."

Lord Flintshire got up and began walking softly up and down the room. In face he was very like his brother, but, though older, he looked younger, for there was a softness about his features extraordinarily youthful.

"As one gets old, my dear Sidney," he said, "one stands in danger of getting old-fashioned. That seems to me to be a very terrible thing. One's own convictions may become hard, fixed in outline, incapable of growth or adaptation, and one may become incapable of imagining that one can be wrong. You may draw your convictions from the highest source: you may be able to say quite honestly, 'I believe with my whole heart that the will of God is so.' But, as Oliver Cromwell once remarked, 'It is just possible that one may be mistaken.'"

He paused a moment.

"I seldom talk so much," he said, "but I have not quite done even now. The younger generation, take them all round, ride life, as I said, on the snaffle. Now, if you choose, you may call that slackness, and as slackness condemn it. But all your condemnation of it will do no good. Martin will continue to be what you call slack; meantime you are in danger of becoming what he would call tiresome. He will also, on occasion, continue to call his sister 'devilish.' Nor is there the slightest reason why he should not. If you or I had called our sisters 'devilish' when we were boys, it would have been undesirable. What you forget is that 'devilish' does not mean now what it meant thirty years ago, nor does Martin mean by it what you mean by it."

Mr. Challoner got up too, his mouth drawn rather tight.

"I am much obliged to you for your advice, Rupert," he said, "but I find I disagree with you in principle so absolutely and fundamentally that there is no use in my discussing with you. I too claim my individual liberty, a very large part of which is concerned with my sense of responsibility for my children."

"My dear fellow, you make a great mistake," said Lord Flintshire.

"I cannot alter my convictions."

"And you will make a great mess of it," said the other.

CHAPTER II.

LADY SUNNINGDALE had few habits and was thus very adaptable, but one was to make a punctual first appearance half an hour before luncheon. But her appearance, though long delayed, was brilliant when it came, and it was as if a fresh and many-coloured sun had arisen to take the shine out of the splendour of the noon-day. Years were the only things in which she was no longer young; but the youthfulness of her mind, tastes, character was perfectly spontaneous and natural, and she still retained to the full all the eager curiosity of youth, all youth's insatiable appetite for pleasure. In person she was very tall and largely made, but she moved with exquisite briskness and vigour, and though stout still clung to her waist. Her hat generally contained a perfect aviary of birds perched about on it, and her dresses to match her tastes were rather youthful in cut and colour. She wore also white satin shoes with extremely high heels, which had been known, when she walked in wet or clayey places, to be drawn with a cloop, like the drawing of a cork, completely off her feet, the heel being driven into the ground by her weight in the manner of a nail. But as a rule she avoided clayey places; indeed, she seldom walked at all, except at this stated time, half an hour before luncheon. But she made up for her lack of walking by talking: this she did on all occasions to as many people as possible, and was extremely entertaining.

She was staying now (she spent the greater part of her life in staying) for a rather extensive week-end, that is to say, from Friday till Monday, with Lord Flintshire, and the morning after her arrival came radiantly downstairs at a quarter-past one. Two irrepressible dachshunds barked excitedly round her, and as she stepped on to the terrace where her host was sitting she was trying, without the least success, to put up a pale blue sunshade with a handle of Saxe china.

"Dear Flints," she cried, "how sweet of you to wait for me! Where is everybody? Yes. Isn't it a divine morning? Everything looks as if it had been washed during the night. Why is one such a fool as ever to leave the country and go to London? If one had a single spark of originality one would never go near it. Yes. Please put up my sunshade for me. I know I look hideous this morning; but it doesn't matter how one looks in the country, which is another of its charms. But I didn't sleep a wink: I never close my eyes in the country; really, London is the only place to live in. I have contradicted myself, have I not? Who cares? I'm sure I don't. Where are the dogs? Please whistle on your fingers, if you can. So piercing, is it not? There they are! Ah, how naughty! Yes, who cares whether one contradicts oneself? It shows, in fact, that one's powers of sympathy and of seeing other points of view are defective, unless one sees both sides of every question, and upholds both vehemently. Yes, do let us walk down the terrace. I adore walking. Oh, Suez Canal, running over the flower-beds like that! How naughty!"

"Suez Canal?" interpolated Lord Flintshire, who, walking by her side, looked like a small rowing-boat towed by a brig in full sail.

"Yes, don't you see how dreadfully long he is? Now tell me all about your brother, who dined here last night. I thought him too fascinating, and we had a great talk about somebody called Kennet, I think he said. Mr. Challoner is very high-church, is he not? His mouth looked to me high-church. There is something perfectly beautiful about high-church mouths. Look at Lady Otterbourne's: her mouth is exactly like your brother's. So is the Bishop of Tavistock's, whom I adore. He plays the flute divinely, looking funnier than anything I ever saw—so funny that I never want to laugh. Somehow, a bishop playing on a flute—or do I mean low-church? I think I must mean low-church. And so your brother is Martin's father. I sent a message by him last night to tell Martin to come and see me this afternoon. I completely lost my heart to Martin last winter. It is terrible to lose one's heart when one is fifty, because one has already lost one's looks, so that it leaves one really denuded. Besides, it seems so careless. That is a chestnut, I think. But everything worth saying has been said years before even I was born. Where is Suez? Naughty!"

Lady Sunningdale's conversation flowed in the manner of a river in flood: it flowed over everything, it foamed and spouted, and there was always the sense—never left unjustified—that there was plenty more to come. It flowed in fact over so many different subjects that her interlocutor had a practically limitless range of topics from which to select the matter of his reply; on the other hand he could fly off on any tangent of his own without initiating incongruity, or again he could be silent, completely confident that Lady Sunningdale would go on. But the last topic suited Lord Flintshire very well.

"Do tell me what you think of Martin," he said.

"But too fascinating, and a genius. That combination is so rare; geniuses are usually quite unpresentable. He was staying with us at Easter and I used to borrow him, as one borrows a book and try to forget to return it. Where is Sahara? Will you whistle again, please? And his playing: well, merely sublime. He can even play Wagner on the piano. Orchestral music on the piano is generally detestable, but Martin—I used to tell him I believed he had instruments concealed about his person. He is quite clever enough too. My dear, you can *hear* the strings. Then he used to draw me caricatures of all the extremely tiresome people that were in the house. And his mimicry! Sunningdale finding fault with the soup, and me telling him he was a gross feeder. My dear Flints, I could have sworn it was us. You know the charming way we behave at dinner. Frank Yorkshire too—you would have thought that nobody could have imitated Frank. But Martin—'Beauty is probably evil in its origin, which accounts for the extreme plainness of good people!' Simply too killing! I suppose your low-church brother doesn't approve of him, or appreciate him. A slight frigidity occurred when I mentioned Martin!"

"He certainly doesn't appreciate all the excellences you have mentioned. I doubt if he really knows they exist."

"That is always the way," said Lady Sunningdale, with a florid gesture of despair. "That very rare product, a natural artistic genius, always makes its wayward appearance in utterly uncongenial places. I am bound to say it usually leaves them before long, but what a waste of time! Dear Flints, don't walk quite so fast, I had no idea this terrace was so interminable. We shall be miles from the house when we reach the end. Where are my angels? But it really is a pity. And I suppose his father will make a curate or a Greek scholar of him."

"That is just what he is afraid he will not do. He was talking to me about it last night."

Lady Sunningdale's attention suddenly and completely wandered.

"You should build a pergola here, Flints," she said. "There is a pergola at Frank Yorkshire's villa in Capri, which is the most divine thing I ever saw, covered with roses. We used to dine there, and earwigs dropped into one's hair, and from the dark one heard those extraordinary Italian melodies from the piazza. That is where I should like to live, to leave the world utterly and entirely and just exist. So unworldly! Yes. My angels, they want their dinner, and so does their mamma."

They had got to the end of the terrace, and Lady Sunningdale gazed about her with roving abstracted eyes. She never did anything, even gaze, without her thoughts being occupied with something totally different; and now as she looked over the great swelling lines of downs which flowed and melted into each other like interlacing muscles away to the horizon across the hollow where the roofs and grey spires of Winchester trembled in a haze of heat, her thoughts were further away than the horizon itself.

"So affected of people to pretend not to like food," she said, "or if it is genuine it shows that they are partly imbecile, lacking the sense of taste. Yes, what Martin wants is to be chucked into an artistic milieu to see what he is really worth. And the artistic milieu is exactly what he hasn't got. He is starving, he is living on himself. Now no artist except the very greatest artist can do that, and even then he dies very quickly. He wants to be soaked and steeped in art. Paris now! There is the artistic milieu there, but the music is generally atrocious, nearly as bad as in London. He could lunch at the Café Champêtre then."

"Why do you wish him to do that?" asked Lord Flintshire.

"Dear Flints, because the cooking is so good. The real artist is a gourmet in everything, including food. Think of the story of Beethoven and the soup. He threw it into the footman's face because it was cold. He could not bear that it should not be hot. Cold soup in one's face—how horrible and, thrown by Beethoven! Even that would not make it pleasant. Certainly Martin has the instincts of a great artist. He has a sense of form in all he does, which I expect means nothing to your brother. Certainly also he has the sense of form in himself. My dear, he is an absolute Adonis, and as slim as asparagus, the English kind."

Lord Flintshire laughed.

"And when do you expect this paragon?" he asked.

"After lunch. To let Martin go on learning Greek and curacies is like looking on at somebody being slowly murdered. Pray do as I tell you and get him away from that terrible Parsonage. Why, the word is enough to upset an artist. It sounds so like parsnips.'

"I feel sure his father would never consent to let him run free in Paris," he said.

"Why not?"

"Because he has the insular distrust of Paris as a residence for the young."

"My dear Flints," she said with some impatience, "if a young man is going to get into messes and make mud-pies, he will make them anywhere. Surely it is the least desirable thing in the world that he should make them in the Parsonage. Yes. You see your brother has so much character himself that he doesn't seriously think that anybody else has got any."

"I wish you would say these things to him," said Lord Flintshire.

"I will if I get an opportunity. But if not Paris, London, Rome, anywhere. Take poor Martin's collar off, and let him roll in the grass. Yes, let us turn. Surely it is lunch-time. But do put up a pergola here all down the terrace, and leave out the earwigs. My angels, we are going to our dinners."

She turned, her very high heels clicking on the hard gravel of the terrace, and paused a moment.

"The mistake in principle which your fascinating brother is making," she said, "lies in thinking that every-one is cast in the same mould, which is his own, and has to be educated in the same manner. Whereas one of the few things of which we can be absolutely certain is that everybody is cast in different moulds. What fools people are really! Fancy trying to make a scholar or a parson of poor Martin! Such a waste too, as well as an impossibility. Sunningdale might as well insist on my taking lessons in juggling or mathematics. Don't you hate conjuring tricks? What is the point of cutting open a loaf of bread and finding a globe of goldfish inside it? Nobody in their senses could call me stupid, you know, but I am morally incapable of adding up three figures correctly. Why? Simply because the process bores me, and I therefore do it wrong."

"That is a fascinating theory of education for the young."

"It may or may not be fascinating, but it is certainly true. The point of education is to develop any taste you may possess, not to bore you with the acquisition of knowledge. Ah, there is Stella Plympton coming to meet us. She has immense charm, and look at the way her head is set on her shoulders. Really, to have a neck is the only thing that matters. A girl with a neck has only to say 'Good morning' for everyone to exclaim 'How brilliant!' Whereas people like me with no neck have to talk from morning till night at the tops of our voices, and wear ridiculous hats, or else everyone says, 'Poor dear, how much she has aged, and how very dull and heavy she is!' Flints, I have immense trials. I often wonder how I keep up as I do, and am so frequently the life and soul of the party. Yes. Everyone made in the same mould indeed! Stella and me, for instance. Flints, your brother is an imbecile. I don't propose to learn Greek because he can talk it in his sleep. Helen too! Is she to be kept in that dreadful Parsonage all her life, and see nobody but district visitors? I think we ought to take your brother's family in hand. He neglects them shamefully; he ought to be prosecuted for criminal neglect. A man has a duty towards his children."

Lord Flintshire laughed.

"And only last night I was telling Sidney that his sense of duty towards them was too strong."

Again Lady Sunningdale's attention rushed headlong away with the bit in its teeth: it was so rapid that one could not say it wandered.

"The last act of the 'Götterdämmerung!'" she exclaimed. "My dear, they gave it superbly the other night, at Covent Garden too, of all places, though the ravens did come in ten bars too soon, and Siegfried had to throw them away. I never sleep for a week afterwards."

The performance in question, therefore, must have taken place at least a week ago, for there was no manner of doubt that when Martin arrived an hour or so after lunch, Lady Sunningdale was snatching a brief interval of much-needed repose after her se'nnight vigil under the cedar on the lawn. The rest of the party, with the exception of Stella Plympton, had dispersed to spend the afternoon in what she considered the violent English fashion: that is to say, Frank Yorkshire and her brother had gone to play golf, Lord Flintshire had taken Lady Sunningdale's daughter for a ride, and Lord Sunningdale himself, who had an insatiable mania for losing large sums of money in what he euphemistically called farming, had gone to feel horses' legs and poke pigs in the back with the Scotch bailiff. Martin, in consequence, who had walked over the field from the terrible Parsonage, and approached his uncle's house from the garden side, found an idyll of placidity occupying the stage below the cedar, for a young woman of about his own age was sitting with an air of extreme content doing nothing whatever, and in a basket-chair close by was Lady Sunningdale recuperating after the "Götterdämmerung." Martin had formed a somewhat copious subject of conversation during lunch, and it required no particular exercise of ingenuity on Stella's part to guess who the tall, straw-hatted figure was. From him again she looked at Lady Sunningdale's slumbers, and, glancing back to Martin, raised her eyebrows as if to ask what had better be done. Then she rose noiselessly from her chair, and beckoning to him with a little amused friendly gesture, walked quietly away from the immediate neighbourhood.

"You must be Mr. Challoner," she said, holding out her hand; "and Lady Sunningdale, apparently exhausted by the prospect of your arrival, is snatching a few moments of repose. What are we to do then? Shall we wake her and risk her immediate displeasure, or let her sleep and risk her ultimate displeasure? We are quite certain to decide wrong."

Much as Martin liked Lady Sunningdale, his instant and instinctive decision was not to wake her, for an enforced *tête-à-tête* with Stella had its obvious attractions. She was nearly as tall as he, and her dark-grey eyes almost on a level with his. Her face was a short oval, slightly and charmingly irregular in feature, the nose a little tip-tilted, the mouth a little full. This, set on the neck, which, according to Lady Sunningdale, could supply the place of intellectual brilliance, made a very good reason for risking the ultimate, not the immediate, displeasure.

"My name is Stella Plympton, by the way," the girl went on. "Pray excuse my introducing so stupid a topic. A person's name matters so very little, does it not? But sometimes it is inconvenient not to know uninteresting things like names and the hours at which trains leave stations. Aren't you thirsty after your walk? Will you not go and forage for fluids? And what are we to do?"

Martin looked at her with his direct lucid gaze.

"No fluid for me, thanks," he said. "What do you advise? One can't go and say, 'Hi, Lady Sunningdale!'"

Stella laughed.

"I couldn't," she said, "but I think you might, if you felt disposed. She adores you, you know."

Martin laughed also, flushing slightly.

"I adore her," he said. "She makes me laugh all the time. And I love laughing."

"So do I," she said. "So please go and say, 'Hi, Lady Sunningdale!' I'm sure it would make me laugh. You won't? Then a false and conventional code of politeness dictates that I should inflict my company on you, though you would probably rather be left alone. Anyhow, do not let us grill here in the sun like beef-steaks. There appear to be chairs in the shade over there. From there too we shall occupy a strategic position, in which to observe Lady Sunningdale's slumbers."

There was a slightly sub-acid flavour about this, of which Martin was just conscious. Stella, it seemed, was conscious of it too, for she explained:

"I feel rather a failure this afternoon," she said, "for Lady Sunningdale asked me to stop and amuse her till you came. The result of my efforts to be entertaining you can see!"

"Please amuse me instead," said Martin.

"I daren't try for fear you should fall asleep too. How is your sister? I remember meeting her once. But though I have never seen you before I feel as if I knew you much better. Really at lunch we talked solidly and exclusively about you. You can do everything, they said, except pass examinations. That seemed to me very admirable; for it is notorious, as Lady Sunningdale said, that any fool can pass examinations. She deduced from that that you can't be a fool."

Martin laughed.

"I ought to apologise then," said he, "though really it isn't my fault that I monopolised the conversation at lunch, or that I am left on your hands now. I hope it wasn't a long lunch."

"Ah, but isn't it the fault of your character that you get talked about?"

"But not that Lady Sunningdale goes to sleep after lunch. At least I don't see how!"

Stella laughed too.

"You put it down to mere lunch?" she said. "But if one was disagreeable one might suggest that it was the conversation at lunch, not lunch itself, that led to the desire for repose. How rude of me!"

Martin looked across to the cedar; he was quite willing that Lady Sunningdale's need for repose should not yet be satisfied.

"But I thought you settled that it was your efforts to amuse her that produced that result," he said.

The sound of Stella's laughter perhaps roused Lady Sunningdale, for she moved in her chair and suddenly sat bolt upright.

"Ah, she is awake," said Stella. "We can pea-shoot each other no longer. What a pity!"

"But that at least is very polite of you," said Martin, rising.

"And that is very modest," she answered. "It might have been true."

Shrill staccato cries came from the cedar as the two walked back across the hot velvet of the lawn.

"Stella, dear, it is too bad of you!" shrieked Lady Sunningdale. "I send for my own particular young man and you monopolise him all the afternoon. Martin, you perfidious monster! What do you mean by flirting with Stella under my very eyes? Did I close them a moment? I think I must have. Is it not tea-time? Where is Sahara? There is a terrible black dog of Flints'. My dear, it is too hot for words. And have you walked all the way from the terrible Parsonage to see me? That is too sweet of you. What have you and Stella been talking about? Stella dearest, if you would whistle three or four times for Sahara. Martin, Frank Yorkshire is here. So odd, two counties in the same house in another county.

Is not geography detestable? Yes. I sat next your father last night. I don't think I ever saw anybody so unlike as you two. I don't think that's grammar. Stella, you went fast asleep I thought in that chair, and when I woke up I found it was me in the other. Where *are* the dogs? Martin, the 'Götterdämmerung' was too exquisite! Ternina! Floods I assure you—I wept floods, and at the critical moment I tugged at my necklace and it broke, and a large pearl fell into the trombone below. Why did you not come up to town, as I told you, for it? Not the pearl; do not be so foolish."

Her slumber had slightly dishevelled Lady Sunningdale, and as she poured forth this surprising nonsense, she effected various small repairs and generally made the crooked straight. Sahara, the delinquent dachshund, recalled by shrill whistling from Stella, waddled pathetically up to her, and a violent wagging of heliotrope in a flower-bed near probably indicated the locality of Suez Canal.

"And we are going to send you to London or Paris or Rome, Martin," she continued. "And we don't quite know which. Tell me, is your father naturally solemn, or is his solemnity beautifully assumed? I don't think anyone could really be as solemn as he appears to be. He sat next me at dinner last night and was quite fascinating. I shall have seven candlesticks on my dressing-table for the future, and be extremely reserved. Dear me, I suppose it would have been better not to have said that. But really his attitude about you is ridiculous. Do imitate him. I am sure you can."

The corners of Martin's mouth quivered slightly.

"I think I won't," he said.

"You mean you can."

"I think perhaps I could," said Martin, guardedly.

"Ah, do. Imitate our conversation last night about matters of high and low church. Wasn't it dreadful? I mixed them up, and I don't know which is which now. Why will Suez Canal always leap about in garden-beds when there is the whole lawn? Naughty! Martin, we have been talking a great deal about you. I am rather bored with you. I stop here over Sunday, and I shall go to church if your father preaches. I think that will give me more influence with him. He said he would very likely come over to tea to-day. I shall never forgive him if he does not, because I want to talk to him about you. We are not going to let you blush unseen any more, and waste your sweetness on the Parsonage air. You've got to go and work. Men must work, though I never saw the slightest need for women to weep. I haven't wept for years, except the other night at the 'Götterdämmerung.' What a charming picture of domestic life, Martin reading Greek history at the table and Mrs. Martin sobbing violently in the corner. Yes! How I run on! I suppose you really ought to go to Germany and eat cherry jam with your chicken."

"How horrible!" said Stella. "Must one take it?"

"If you want to enter into the essential Teutonic spirit you must. You might as well hope to feel like an Anglo-Saxon without being always in a rage or playing violent games, as try to be German without jam. How I hate women who play games! They are nearly as odious as men who don't. Let us go indoors, and Martin shall play to us till tea-time. Afterwards he shall play till dinner-time."

Lady Sunningdale surged slowly to her feet and looked helplessly about.

"Where are the dogs?" she said. "It is too tiresome:

they are sure to stray into the woods, and Flints' horrid pheasants will peck them. My darlings! ah, there they are, amid what was once begonias. It looks more like a battlefield now; how naughty! Come at once, all of you!"

There was no doubt whatever that Martin's piano-playing was of a very remarkable order, and before he was half way through Chopin's first *ballade,* Stella, who had been accustomed to consider the piano as an instrument for the encouragement of conversation after dinner, or at the most as the introduction to the vocal part of a concert, found herself sitting bolt upright in her chair with a strange tingling excitement spreading through her and a heightened and quickened beating of the blood. She was essentially unmusical, but something in this was extraordinarily arresting: her nerves, if not her sense of melody, were at attention. As for Lady Sunningdale, she always gasped when Martin played, and did so now.

"Too heavenly!" she said at the end. "Now make me miserable. Play the rain on the roof. Tum, tum, tum, tum, don't you know. Yes, how clever of you to guess!"

It was rather clever, for Lady Sunningdale's rendering did not really resemble any one tune in the world more than any other.

Martin paused a moment. Then the slow sullen drip of hot steady rain on the roof began, as it sounded to a man who was left alone in an alien land. It fell with hopeless regular iteration from grey skies; then there was the gurgle of some choked gutter, and the collected water overflowed and was spilt with a little chuckle. Very distantly on the horizon remote lightning winked and flickered, but there was as yet no sound of thunder in the dark sultriness of the afternoon, but only the endless monotonous rhythm of the dropping rain. Then faintly at first but with slow crescendo there were heard the distant drums of thunder, buffeting and rumbling among the hills. Then all at once the rain grew heavier, larger drops as if of lead fell beating with a resonant insistence on the roof, and the voice of the storm grew angry and articulate. Suddenly, with an appalling crash, it burst immediately overhead, drowning for a moment the beat of the rain, and by the blaze of the simultaneous flash, sea, sky, and the wave-beaten rocks of Majorca leapt into light. Then, as thunder will, it drew away, and for a time the rain was not so heavy; but again the storm swept up, once more the chariots of God crashed on their way above them, the wild lantern of the lightning flared this way and that, and once more again after that stupendous riot in the skies, the hot darkness was punctuated by the dreadful melancholy of the dripping rain. Then the storm growled itself away into the distance; a little light came back into the weeping skies, the pulse of the rain grew fainter, and again a choked gutter gurgled and overflowed. Suddenly through some unconjectured rift in the clouds, one beam of the sun, divinely clear, shot down for a moment on them with excellent brightness. Yet it was only for a moment; again the clouds drifted up, and the rain, which for that minute had ceased, began again, dripping with hopeless regular iteration on to the roof, as evening closed in, some evening far away, in a land of exile, beneath an alien sky.

Effusive as she usually was, and accustomed to fill any interval of silence that might conceivably occur, with discursive volubility, even Lady Sunningdale was silent except for an "Oh, Martin!" which she no more than whispered. For there was that in the room which, in spite of her superficial frivolity and the dragonfly dartings of her mind, she knew and recognised and adored, that true touch of Art which makes even of things that are common and unclean, gems and jewels. Stella too said nothing, but sat still, much more upright than her lolling wont, holding the arms of her chair. From where she sat she could see Martin's profile cut with great clearness of outline against a brocaded screen of scarlet and gold that stood beyond the piano, and between the music and the musician she was dumb. Even in the desultory accidental conversation which she had had with him during the slumbers of Lady Sunningdale, there had been something arresting to her in his brilliant boyish personality; and now from his finger-tips there flowed out, so it seemed to her, a personality just as brilliant, but either very mature, or by the instinct of genius still boyish, but clad as it were in the purple of the artistic nature. There was nothing amateur about it; and unmusical as she was, she could not help recognising the certainty of the performance.

For a few moments after the last note had died into silence he sat silent also, with head bent over the keys. Then he looked up.

"Is that enough, Lady Sunningdale?" he asked.

"No, you angel from heaven, it is never enough!" she cried; "but play something different, something brilliant —I should expire with several hollow churchyard groans if you played that again. It makes me miserable. Play something *virtuoso,* and let me come closer, where I can see your hands."

She moved to a low chair to the right of the piano.

"Brahms; Paganini variations," he suggested.

"Ah, yes, do. It makes me shriek with laughter."

Then, with the same absolute facility and certainty,

with the same cleanness and perfection, suggesting indeed a slim poised figure, he took a header into that ridiculous theme. But out of the foam and bubble beneath his hands flowers grew, stars were scattered, and all nature went mad with dancing. But when the riot of jubilance was at its height a tall, severe figure suddenly appeared at the French window of the drawing-room, advanced very audibly on the bare boards, and spoke sufficiently loud to be heard.

"Ah, Lady Sunningdale," said Mr. Challoner, "how are you? And Martin wasting his time at the piano as usual. How kind of you to let him play to you!"

Martin wasted no more time there; at the noise of interruption, before his brain had conjectured who it was, his hands stopped, the eager active vitality died out of his face, as when a candle is blown out, and he banged a random chord in sheer rage. Then instantaneously he recognised the voice, and he rose quickly from the music-stool, trembling.

"Yes, wasting my time as usual," he said excitedly, the artist in him suddenly struck dead, leaving just an angry, startled boy. "I must go home, Lady Sunningdale. Thank you so much for letting me play to you, and I hope I haven't bored you. Good-bye. I have a lot of work to do."

He closed the piano lid as he spoke, but it slipped from his fingers and shut with a bang that set all the strings jarring.

"Ah, how could you interrupt like that?" cried Lady Sunningdale to his father. "Yes, how are you, Mr. Challoner? Martin, pray begin it again. We will all sit quite quiet, without stirring a finger or breathing. You are superb!"

His father sat down, distressed at Martin's rudeness, but honestly desirous of being sympathetic.

"Dear boy, I am so sorry," he said. "Pray play your piece!"

"I can't," said Martin. "I don't know it."

For a moment father and son looked at each other, the one with surprise and indignation, the other in impetuous rebellion and anger.

"Lady Sunningdale asks you to play again what you were playing," said his father, the desire to be sympathetic vanishing, the sternness deserved by this deplorable lack of manners in Martin increasing every moment.

"It is quite impossible that I should play it," said Martin. "I couldn't play a note of it."

"You seemed to me to know it," said Mr. Challoner. "Surely you have played it a hundred times at home."

Martin was really incapable, in the shock of this transition from the world which he loved, and in which he was at home, to this other world, of decent behaviour.

"More like a thousand times," he said, and simply and directly left the room.

There was a somewhat awkward pause. Mr. Challoner was seriously angry with his ill-behaved son; Lady Sunningdale was disgusted at being deprived of her music; and Stella, with a natural eye for drama, was immensely interested. It seemed to her there might be a good deal of drama behind this little incident. Then, luckily perhaps, Lady Sunningdale remembered that she was, so to speak, on a mission to the dark ignorance of Mr. Challoner, that savage in matters of art, on behalf of Martin, and she put her disgust in her pocket.

"It is delightful of you to have come over to see me," she said to him, with her easy natural charm. "Yes, I suppose Martin wastes a terrible lot of time at the piano when he should be doing Greek history. Demosthenes! How fascinating! Stella, dearest, do see what Suez Canal is doing, and slap him. And will you tell us when tea is ready? But do you know, Mr. Challoner, Martin plays remarkably, really remarkably?"

Stella, as she was meant to do, strolled out through the window by which catastrophe had entered, leaving the two others alone.

"Yes, it is that incessant waste of time that distresses me," said Mr. Challoner. "But the piano at the Parsonage is so old that he hardly cares to play on it. But first, I must apologise to you, Lady Sunningdale, for the extremely rude way in which Martin behaved to you. I promise you he shall make his apologies in person!"

For a moment her irritation mastered her.

"He apologise?" she cried. "It ought to be you. Dear Mr. Challoner, how rude I am, pray forgive me. But you don't know, you can't know, what music is to Martin. You don't know what divine glorious mood in him you shattered. It was like throwing a brick at an iridescent soap-bubble. I suppose Brahms is a name to you like Smith or Jones."

Then she recalled diplomacy again.

"So difficult to understand Brahms, is it not?" she said. "That is the fascination of it. But I assure you it is worth thinking over—Martin is wonderful. He has improved so enormously too. He is not second-rate or third-rate, but first-rate. What have you been doing to him?"

"You mean at playing the piano?" asked Mr. Challoner, as if he had said, "sweeping a crossing."

Lady Sunningdale longed for Sahara to bite him.

"Yes, at playing the piano," she said, swallowing her irritation again. "He ought to study, you know. He is wasting his time, that is quite true, but not at the piano. I am dreadfully impertinent, am I not? But Flints is an old friend and Martin is his nephew, and music is music, so I feel it very strongly. Of course, it is only natural that you, Mr. Challoner, with your earnest nature and your serious aims, and all that—you were too interesting last night; I lay awake for hours thinking over what you had said—should consider poor Martin very frivolous, but he is an artist to his finger-tips. It is his nature. Mon Dieu! what finger-tips too! You know he was playing, and playing I assure you with consummate ease when you interrup—when you came in, a thing that really great pianists require to practise for months!"

"You are too kind to take such an interest in my lazy son," said Mr. Challoner, still very stiffly, so stiffly, in fact, that Lady Sunningdale looked hastily at the fireplace, thinking he must have swallowed the tongs.

"I assure you it is not kindness that prompts me at all," she said. "It is mere justice and mere economy; I am very economical—ask Sunningdale. The world cannot afford to lose a talent like that. If he is like that when he is practically uneducated, to what may he not grow? Heaven knows, the world is so very stupid that we should hoard and save every grain of talent that exists. It is like what you so beautifully said to me last night about the ten talents in a napkin."

"Surely not," said Mr. Challoner, a faint smile breaking his gravity.

"Well, the one talent then; I have no head for numbers. And poor Martin's talent seems to me to be put in a very damp napkin, except now and then when somebody like me lifts up a corner of it and lets the sparkle of gold appear."

It happened very rarely that Lady Sunningdale was stirred into such coherence and earnestness. As a rule her multifarious little interests were like children playing "King of the Castle," rapidly pulling each other down from their momentary pre-eminence, first one and then another perching precariously on the summit. But certainly the most long-lived "king" there was music, and Martin's future, with the rain-storm of Chopin and the mad frolic of Brahms still in her ears, was very securely throned.

"Think me impertinent, my dear Mr. Challoner," she went on. "Think me what you will, only do give your most serious attention to what I say. Martin devoting his fingers, his brain, the power of his extraordinary artistic nature to ancient history, is a thing to make Julius Cæsar weep. The pity of it when he might be starting us all on a new chapter in music! Really I believe that to be possible. And really I am in earnest; and when, as I hope, you know me better, and see how completely scatter-brained I usually am, you will appreciate how deeply I feel this!"

"You mean that my son should devote the most useful, the most active years of his life to playing the piano?" he asked.

"Playing the piano?" she cried, feeling it was almost hopeless to try to make him understand. "That is, of course, a thread in the golden garment of music; but to take piano-playing as synonymous with music would be the same thing as calling the baptism of those of riper years the same thing as Christianity. Music—music, that must be his life. Flints told me this morning that you found him slack, lazy. So would you be if you had to learn scales, just as he may be—I am sure he is—at classical studies."

"What do you propose then?" he asked, inwardly rather rebelling at the consideration he felt somehow forced to give to her eagerness. For in spite of her discursiveness it was so clearly impossible not to recognise the surprising quickness and intuition of her mental processes.

"Why, just what I have been telling you. First let him throw his dictionaries and histories into the fire."

"I have an immense, a vital belief in the educating power of the classics," said Mr. Challoner.

"For everybody? You cannot mean it! Can you tell from looking at the picture if the artist knew Latin? Or pick me a piece of Greek out of 'Tristan and Isolde?' In any case Martin has spent some ten years at them, he tells me, and what is the result? He fails to pass his examinations. Whether they are a criterion of education, or whether they are an instrument, he or they have failed. He is second-rate at that, third-rate—it is all one. There is first-rate, and—the rest of the world. What is the good of turning another second-rate person into the sheepfold of the second-rate, particularly when on other lines that person has all the appearance anyhow of being first-rate? Well, that is what I think. How kind of you to let me talk so! Where are my angels? Is it not tea-time?"

Lady Sunningdale's unparalleled effort in concentration of thought here broke completely down, and a whole tribe of clamouring competitors invaded the castle of her mind, dethroning the "king."

"Yes, Martin really was playing too divinely," was the "king's" expiring cry. "So like a great artist, too, to bang down the piano lid when he was interrupted! Beethoven did it too, you know, and shouted, 'I play no

4*

more to such swine.' So delicious of him! And Helen, how is she? You must bring her over. Frank Yorkshire is dying, if not dead, to see her. He is one of those people, you know, who does nothing and appreciates so much. So infinitely better than doing a great deal rather badly, and not recognising the first-rate when you see it. And are you going to preach on Sunday? I should have been so happy if I had been a man to have lived in a country place like this and just spent my days in doing a little good among these simple people. How beautiful it must be! I abhor London, so shallow. Yes. You really must preach on Sunday, Mr. Challoner, otherwise I shall stay at home, and read improper novels. You would not like to have that on your conscience, would you? People are growing terribly slack about Sunday, are they not? Yes, shall we try to find some tea? Talking makes one so hungry."

CHAPTER III.

Mr. Challoner was seated at the very orderly table in his study, on which, neatly corrected, revised and arranged, were the sheets of his sermon for the next Sunday. In front of him, with his face towards the window, stood Martin. Neither father nor son wore a very pleasant expression: Martin looked like some timid wild animal, at bay in a corner, frightened into a sort of desperation, while his father's thick bushy eyebrows were contracted into a very heavy frown, and his mouth was tightly compressed, as if he was holding back with difficulty some impulse of anger that nearly mastered him.

"I was ashamed of you," he said; "I was ashamed that a son of mine could behave with such abominable rudeness to Lady Sunningdale and me. A few years ago, when such behaviour would have been more excusable because you were younger, I should have given you a whipping!"

"I am sure you would," said Martin.

Mr. Challoner's face grew a shade paler.

"Martin, I wish you to understand once and for all," he said, "that I will be treated by you, both in public and in private, with ordinary respect and courtesy."

"I have already told you I was sorry I was rude to you," said Martin, speaking very quickly and incisively, with an odd little tremor of angry fright in his voice.

"You have often told me you were sorry lately," said his father; "and almost before the words were out of your mouth I have had occasion to find fault with you for something else."

Martin gave a short, mirthless laugh.

"That is quite true," he said; "I can't do right, it appears."

Mr. Challoner paused a moment: Martin had never before come to open words with him like this.

"What do you mean by speaking to me like that?" he asked, in a voice scarcely audible.

There was no answer.

"I have asked you a question, Martin," he said, his voice rising suddenly.

Martin pushed back his hair with a hopeless gesture.

"What answer do you expect me to give?" he asked impatiently. "There is no answer to such a question. You get angry with me and you frighten me; I think you do it on purpose. You have frightened me into silence all my life, now you have frightened me at last into answering you. I hate anger, it makes me sick. And you have been angry with me every day since I came home from Cambridge."

He sat down on a chair behind him with a sort of dull indifferent acquiescence in whatever might happen, his face sullen, frightened, joyless. It seemed as if it could scarcely be the same radiant boy who had played Brahms an hour ago.

There was a pause, and all the imprisoned longing for love in the father beat dismally at its bars, for he felt, and felt truly, that just now Martin almost hated him. It seemed terribly hard that his own daily and constant desire that Martin should grow up a useful God-fearing man, industrious and earnest, should be the bar that separated them; yet so he knew it to be. Had he been a weak indulgent father, one who had not implanted in him the unbending, ineradicable sense of his duty towards the son whom God had given him, how sweet might have been the human relations between them! His love for his son was the very reason why he corrected him, that and the duty attached to his own fatherhood; and when he saw him slack, lazy, or, as now, wanting in courtesy and respect, it was still from sheer duty that his anger sprang. And now, for the first time from Martin's own lips, he heard the effect. He frightened him, on purpose, so it appeared. Was this then one of the hopeless, incomprehensible puzzles that God seems sometimes to set His groping children, this fight between duty and love, in which one must lose and be vanquished? It seemed to him cruelly hard if this was so.

Martin felt his mouth go suddenly dry as he spoke, but he was past really caring what might happen. His father, he knew, was about as angry with him as he could be; and he himself hated and feared his anger in the instinctive unreasoning way in which a grown man will fear something which can really hurt him no longer, but which he feared in childhood. That vibrating note was in his father's voice which he associated with early failures of his own with Latin declensions, and the hint of what would have happened to him if he had been younger also carried him back to early dreadful scenes. But, finding his father did not reply, he looked up at him and saw that the anger in his face had been extinguished like a wind-blown lamp. But all tenderness, all sense of being intimate with him, was so alien to Martin that he did not trouble to guess what emotion had taken the place of anger. Anger, however, was gone, taking his own fear with it; and, with a certain mercilessness characteristic of youth, he deliberately, so to speak, hit back.

"Whatever I do you find fault with," he said. "I try to please you; it is no use. Would it not be better if I went away? There is no good in my stopping here: I don't suppose this sort of thing gives you any pleasure. Uncle Rupert, I am sure, would let me go and stay with him in London next week till the Long Term begins at Cambridge. That will be in another fortnight. You told me you wished me to be up there all the time. So would it not be much better if I went away?"

His father did not reply at once, but sat fingering his writing things with rather tremulous hands.

"Are you not happy at home?" he asked at length.

"No," said Martin shortly.

The brevity and certainty of this struck more deeply yet. If Martin a few moments before had felt sick at his father's anger, the latter was certainly more to be pitied, now.

"Martin, what is the matter between us?" he said.

"I don't know; but it's the same as it has always been, only it's rather worse. I can't please you, I suppose, and you are always down on me for something. It is to be hoped it is doing some good, because otherwise it seems —well, rather unnecessarily unpleasant. First it was my work, then what I said to Helen, then the 'Mill on the Floss,' and now this. To-morrow it will be something else. There is sure to be something. I daresay I don't understand you, and I know you don't understand me. This afternoon, for instance. Oh, it's no use trying to explain," he said.

"It may be the utmost use. It may make the greatest difference. I only wish that you had said to me years ago what you are saying now. I have tried to be a good father to you, but sometimes, often, I have been puzzled as to what to do. You don't confide in me, you don't tell me your joys and pleasures, and let me share them. I often hear you laughing when I am not with you. But when I am, not so often."

Martin half shrugged his shoulders as if to say, "There we are again."

"That is quite true," he said. "But what can I do when music, which to me is the greatest joy and pleasure in life, seems to you just a waste of time? You have often told me so. You don't know one bit what it means to me, and as it seems to you a waste of time, how can I confide in you about anything you don't really approve of and of which, you will pardon me, you are absolutely ignorant? In the middle of the Brahms or whatever it was, you come in and interrupt by saying that I am wasting my time as usual. I might as well come in in the middle of prayers and say you were wasting—— There

I go again. I am sorry. That will show you how hopeless it all is."

Mr. Challoner was silent a moment, really too much pained to speak. But he was wise enough to recognise that to say anything just then would be to effectually stop the only confidence that Martin had reposed in him for years.

"Well, Martin," he said, after a moment.

"Ah, it's no use," he said. "Even at the very instant when I am consciously trying to be careful, I say something like that, and you are shocked at it. But I meant it; it exactly expressed what I meant. Music is to me like that. You never thought that possible. All these years you have been thinking that I was very fond of music—just that—and wasted a great deal of time at the piano. Whereas it seems to me that I am wasting time when I am reading 'Thucydides.'"

"That is what Lady Sunningdale said. She talked to me about it after you went away. You know her well, do you not?"

"Yes; she has been tremendously kind to me."

His father rose.

"You must go now, dear lad," he said. "I have got some work to do before to-morrow. And let us try, both of us, to find more of a friend in each other. I shall never have another son, and you will never have another father. It would be very sad, would it not, if we did not, each of us, make the best of that relation?"

There came into his beautiful brown eyes the shadow of tears, and Martin wondered.

"I will try, father," he said.

Mr. Challoner did not at once begin the work which he wished to finish before bedtime when Martin left him, but sat with his head resting on his hand, thinking very deeply. He was much troubled and perplexed, and his future line of action, usually so clear to him, so precisely indicated by his sense of duty, and, to do him justice, so undeviatingly followed, was now very misty and ill-defined. Hitherto he had never entertained any serious doubts that he was not doing the best possible for Martin, both in always correcting and admonishing when he seemed to be idle even in trifles where some small carelessness on his part indicated the danger of his falling into slack or slovenly habits, and in his conviction that school and college education in classical subjects was the best possible method of training and developing his mind. He did not in the least even now with regard to the latter think it certain that he was mistaken, but it had been brought home to him very clearly in the last twenty-four hours that other people thought he was. For his brother's opinion he always felt a considerable respect, but for Lady Sunningdale's, though he wondered at it, he could not help feeling more. A dozen times yesterday at dinner, a dozen times more this afternoon, he had asked himself how the observations of a woman who really appeared to be scarcely capable of consecutive orderly thought could be worth consideration, but as often some plump grain of solid sense, showing acuteness and perception amid the husks and chaff, answered the question. He himself was conscious of not being quite at his ease with her, but he could not help admiring her intense vitality, her speed, her busy, acute inquisitiveness. And it was she who hailed Martin, poor desultory, idle Martin, as a genius.

Suppose he took their advice and let his son go free into that world of which he himself knew so little, of which, however, he had so abundant a mistrust, how dangerous and hazardous an experiment! Martin, with his slackness, his ineradicable tendency to what was easy and pleasant, Martin, above all, with this apparently so great musical gift, unsuspected by his father, but adored by others, was exactly the sort of boy to be petted, spoiled, ruined by the careless highly-coloured butterflies which Mr. Challoner believed to dance there all day in the sun. To them, music, painting, drama, the visible arts, were ends in themselves, the object being enjoyment; while to him such a doctrine savoured almost of profanity. To him, painting, sculpture, music were recreations which might at intervals be innocently allowed to the earnest worker; but even in such times of refreshment the Christian would look for something more, and find in beauty that which should lead his thoughts to the Fountain and Creator of it. Such, however, was not the view, as he was aware, of the world of art into which he was invited to let Martin plunge; to them music was sweet sound and led the soul nowhere but to music; painting was line and colour; sculpture was form, and the end and fulfilment and consummation of it was perfection of form and the appreciation thereof. About this latter branch of art he had never been able to come to a definite conclusion. Certainly studies in the nude seemed to him to be things dangerous if not inherently sensual.

"All art is perfectly useless;" he remembered having read that sentence in some book of Martin's which he had found lying about. A rapid glance at it on that occasion had justified its confiscation, and a few words to Martin on the subject. But that sentence occurred to him again now, for there in half a line was expressed tritely and unmistakably the exact opposite of what he held to be the truth. All art, he would have said himself, that does not—apart from the natural and innocent enjoyment of it—raise and elevate the soul, is not art at all. As a corollary, the highest form of painting in his eyes was religious painting, because it led by a direct road to its goal, the highest form of music, religious music. These two were wholly laudable: Raphael, so to speak, shook hands with missionaries, and Handel took Luther's arm. But at the other end of the line of artists came those who, however consummate was their art, treated of themes which in themselves were dangerous or, worst of all, by clothing sin in melodious and beautiful garb, rendered it, even if not attractive, at any rate more venial. He himself, as has been seen, was not musical; but when a few weeks ago he had found himself in London with Martin, and, with the eminently laudable desire of getting more into sympathy with his son, had taken him to see "Tannhäuser" at the opera, the evening had not been wholly a success, for the curtain had not risen ten minutes on Venusberg before his incredulous horror had deepened into certainty, and he had got swiftly up and peremptorily ordered Martin to leave also. And Wagner was hustled by him into the outer darkness, to gnash teeth in company with Zola, George Eliot and Titian.

Here then is stated in brief, so that the real and soul-searching difficulty in his course of action with regard to Martin's future may be better understood, the attitude of Mr. Challoner towards art. With the whole force of his strenuous, upright soul he believed that one thing in the world alone mattered, and that Art, Science, Knowledge, were at the best but by-paths that led on to the great high-road of the Gospel. In that they contained many things of beauty the worker was allowed to wander in their coolnesses at times for the refreshment of his weariness, but all the beauty he found there was but the sign-post pointing him back to the highroad. Other by-paths were there also, beautiful as these, if one looked on the outward form only, but instinct with danger, and of an evil glitter. Such led through tangled gardens of meretricious gaudiness, but if one stooped to pluck those poisonous flowers they were vitriol to the fingers; and the unnameable beasts of darkness, coiled among the leaves, alert and ready to spring, would fasten on the hand.

Martin had left his father's presence that evening with an idea that was really quite new to him. The truism in fact that a father loved his son had suddenly emerged from those dull ranks, and taken its place in the far more notable array of truths. For the interview which had begun in a manner so dismally familiar to him, except that in this case it was set one or two octaves higher than usual, had ended in a manner unexpected and unprecedented. Never before had he known, though he had vaguely taken it for granted, that his father really cared for intimacy and love in his relations with himself. At any rate he had never seen the fact bare and exposed, for whenever it had shown itself it had always been wrapped up, so to speak, in the memory of some rebuke. But to-night it had flashed on him; he had seen through these coverings, and a heart of gold shone and beat within. And, with the natural instinctive generosity of youth, he himself was quick to respond; and though his habitual reserve and shyness with his father could not at once be dispersed, so as to allow him any effusive rejoinder, his response had been very genuine, and his resolve, as he left the study, to explore and develop the reef which had suddenly gleamed in what, to be frank, he had considered hard, unyielding rock, very vivid. With this in his head, ready to be matured by the unconscious processes of sleep, during which the brain, though the senses lie dormant, goes on delving in its difficulties and groping for light, he went up to bed.

As he undressed, his mind flashed quickly backwards and forwards through the events of the day; for a moment a smile uncurled his lips as he thought of some extravagance or incoherence of Lady Sunningdale's, the next his mouth was pursed again into a low whistle of some half-a-dozen bars of a tune that ran in his head. That Brahms—to which had come so fruitful an interruption—what a delicious piece of boisterous irresponsibility! It had infected Stella Plympton too: he had known that from a glance at her wide eyes and half-opened mouth when he began. Then suddenly, just before the interruption came, she had given one heavenly ripple of unconscious laughter at some surprising piece of *virtuosité*. Yes, she understood, understood probably better than Lady Sunningdale, who always gasped. The gasp, it is true, was a great compliment to his nimble fingers; but it should be as impossible to think of fingers or nimbleness when that was going on, as to think about the chemical constituents of water when one is satisfying a noble thirst. Then came that dreadful scene in the study, with its utterly unexpected end. Well, he would try, anyhow.

The moon was shining outside against the blind with an amazing white brilliance, and as he undressed he went across to the open window and let the flood of cool light shine in. It made the yellow flame of his bedroom candle look insufferably vulgar and tawdry, and blowing it out he again crossed to the window and sat there while the stirring of some fragrant breeze sent its soft ripples against his skin. As Lady Sunningdale had said, he was a gourmet in sensations, and the exquisiteness of the sleeping summer night, peopled with ivory lights and ebony shadows, and the great velvet vault of the sky pricked by the thin remote fires of innumerable stars, and lit by that glorious sexless flame of the moon, smote him with a sudden pang of pleasure. Somehow all this must be translatable into music: the stars scattered over the sky were like staccato notes of strings across the great tune of the moonshine; it was the first slow movement in the great symphony of night and day. At sunrise the scherzo would laugh and dance down the breeze of morning with a thousand quivering leaves and a million nodding flowers, trees waving, birds among the branches. Noonday would combine all the powers of light and air into a third movement of intolerable splendour. . . .

He got up from where he sat, and stretched his arms wide, as if to embrace it all. Then, half-laughing at himself, he dived into his nightshirt, leaving the rest of his clothes in a heap on the floor, and, as his custom always was, laid his face on his hand and fell asleep.

It was still early when he woke, but the sun was up, and even as he had anticipated before he went to sleep, the slow movement of the moon had given place to a dancing rapturous scherzo. A breeze stirred with a short sweeping rhythm among the trees, birds chirped in the leafy temples, and the sparkle of the early sunlight gave an inimitable briskness to the young day. Then with a sudden ebb in the full tide of his joy of life came the thought that it was Sunday, a day in that house neither of rest or gladness in his view, but one much taken up with lengthy unmusical services, in which there was a great deal of singing, with intervals in which no amusement could be indulged in.

He walked from his window back to his bed and looked at his watch. It was still not yet seven, but the "land of counterpane" was no longer desirable or even possible, and putting on coat and trousers he went quietly downstairs and out across the lawn into the fields beyond, where a bathing-place had been scooped out of the river-bed. Till breakfast, at any rate—still two hours away—he need put no restraint on the flood of vitality and joy that ran this morning in spate through him and this beautiful world. There were two hours of it, with the cool shock of the racing water, the caress of the warm wind, the sense of being alone, and young, and out-of-doors. Pagan it might be, but irresistibly delightful.

Then suddenly while still thrilling with these joys the mellow tones of the church-bell struck across the staccato sounds of life, and all at once the scene with his father the evening before, and his own resolve to try to please him, flashed into his mind. The bell, he knew, must be for the early celebration in the parish church, and he had still twenty minutes, enough, if he was quick, in which to dress in the prescribed Sunday garb (though why black was suitable to Sunday he had long given up trying to guess, leaving it to rot away among the unconjecturable riddles of life), and, a thing which pleased his father so intensely, play the hymn on the melancholy one-manualled organ, the curious quavering tones of which formed so remarkable a contrast to the nasal tones of village voices. So, with something of a sigh for his renunciation of the river-bank, he hurried back home, and before the bell had ceased ringing passed through the churchyard, where yew-trees of noble growth looked down upon the horrors of the modern stonemason, with his "chaste" designs and "handsome" crosses, into the grey cool church.

To judge by the interior it is probable that the mouth which Lady Sunningdale so much admired in the vicar and the Bishop of Tavistock was a low-church mouth, for Mr. Challoner at any rate did not attempt to make any appeal to the souls of his parishioners by means of the senses. Two brass flower vases of that curiously feeble design that somehow suggests at once low-church ecclesiasticism stood on the altar, over which a flood of mauve and magenta light poured in through misshapen figures of apostles and prophets in the east window. In one transept stood the organ, to which Martin directed his steps, the pipes of which, framed in a wooden border ornamented with fretsaw work, were painted white, with a scroll of red pattern in line embellishing their top ends. Behind the organ-bench was a red plush curtain with golden fleurs-de-lys stamped on it, to screen the person of the organist from the eyes of the congregation. The seats for the people, who were thinly scattered over the church, faced eastwards, and were made of shiny varnished pitch-pine, while the floor of the aisles and accesses was tiled with a cheerful ecclesiastical pattern in violent blue and Indian-red, and pierced here and there with gratings of cast-iron work, through which in winter came the hot stale blasts from the warming apparatus. A black iron stove stood near the font at the west end of the church, and rows of somewhat dilapidated rush-bottomed chairs denoted the place allotted to the school children.

To Martin, who for the last two months had been accustomed to the grey dimness and carved spaciousness of King's Chapel, the first sight of these staring crudenesses came with a shock of almost physical repulsion. Why had it been done? What did it all mean? What emotions were the ill-coloured, badly designed windows intended to arouse, or what was the affinity between pitch-pine and worship? Impressionable and impatient as he always was, he nearly turned back after he had opened the door and was confronted by this half-forgotten tawdry brilliance. Then the motive which had made him forsake the cool riverside, the desire to please his father, prevailed.

The organ was blown by a small boy with a highly-polished face who stood directly by the player's left hand, and, since the bellows were not powerful enough to supply the lungs of the organ unless plied by an energetic arm, was often blown too, and breathed heavily into the organist's ear. It was still a few minutes to eight when Martin came in, and found the village schoolmaster preparing to begin that series of somewhat elementary harmonies to which is given the vague title of a "voluntary." But he slid quickly off the seat with a smile of welcome to the other, and in a searching whisper told him what the hymns were going to be, and what "Kyrie" would be sung between the Commandments. This latter information was given with a self-depreciatory blush, for Mr. Milton was not at all mute and inglorious, but composed chants and hymn tunes with so many accidentals that the choir quailed before them, and garnished them with accidents.

Martin glanced at the organ-stops; these were "Bourdon" (which sounded as if you were playing pedals when you were not, and was much in request), "Open Diapason," "Flute," "Cor Anglais," and a few others of more doubtful import. He added "Tremolo" to certain other soft stops, in curiosity as to what it meant, and began the first bar of the prelude to "Lohengrin." But as "Tremolo" seemed to convert other sounds into a distant bleating of sheep, he hastily put it in. Five minutes later the vestry door in the transept opposite opened and the curate followed by his father came out. Mr. Challoner looked up as he entered, saw Martin's head above the curtains of the organ, and a sudden warm tide of thankfulness and love glowed in his heart. Surely the dear lad could not go very far wrong, if he sought strength here.

The worshippers were but few, and it was not long before Martin was out in the sunshine again, but with all the joy and exhilaration of the earlier hour by the river driven out of him. Like most very emotional people, religion was as essential to him as breathing; but in him it was a natural child-like religion, that springs primarily from the huge enjoyment of the beautiful things in this world, for which he had to thank somebody. And though it would be impossible to say that it was not real to him, yet a London fog, so to speak, would make a pagan of him for the time being. And now, though he did believe in the truth and reality of the service in which he had taken part, the deadly ugliness of the church, the melancholy voices, Mr. Milton's "Kyrie" ten times repeated, the intolerable voices singing absurd tunes, had risen like a London fog between him and it. The service had passed over his head like a flight of birds unseen in this dreadful atmosphere; he had heard only the rustle of their wings. But what he had been conscious of with every jarred fibre in his being was the gross material ugliness of the sights and sounds of this last hour. Why should "throne" be allowed to rhyme with "join" in sacred subjects, whereas it would be admissible in no other class of poetry? Was it because anything was good enough in a hymn, or because those who were responsible for the "form" of English worship were entirely without any sense of "form" themselves? Or why in church allow music that would be tolerated nowhere else? Or why have windows in the House of God, which for colouring and design could only be paralleled in the worst type of suburban villa? Pitch-pine seats, tiles again only to be found in the fire-places of villas and the aisles of churches? Often before, though never perhaps so vividly, had the ugliness of Protestantism struck him; often before, though never perhaps so insistently, had his nature, wishing to aspire, demanded beauty as its ladder. Most of all here was beauty necessary, for the sublimest act of all was here performed, the worship and praise of God, the sacramental approach to Him. Even as a little thing, a little rhythmical noise, may utterly distract a man's attention from a subject which requires concentration, so this ambient ugliness utterly distracted Martin. Only ugliness was no little thing to him.

He had not long to wait for his father, for he followed him almost immediately out of the vestry, and his face lit up with extraordinary pleasure when he saw that Martin had waited for him. Here was his highest joy: to see his children with him in that divine act, and find them caring, lingering for him; and the consciousness of that compact the night before was as vividly present in his mind as it had been in Martin's when he left the delights of the river-bank at the sound of the church-bell.

"Dear lad," he said, "the first thing I saw when I came into church was you, and I was so thankful."

Then, with the active desire to get into Martin's sympathy, he went on:

"And what was that beautiful, exquisite tune you played before service?"

Martin brightened.

"Ah, I am glad you liked it," he said cordially. "Is it not beautiful? It was Wagner, the beginning of the overture to 'Lohengrin.'"

Mr. Challoner's face grew suddenly grave. Wagner was identified with "Tannhäuser" to him.

"Certainly it was most beautiful," he said; "but do you think it quite—quite suitable to play something from an opera in church, before the Holy Communion, too? One wants everything, is it not so, to be of the highest?"

Mr. Milton's "Kyrie" occurred to Martin, but he dismissed it.

"I don't see why one shouldn't play an opera overture, father," he said. "Does not the fact that it is beautiful make it suitable?"

"But the associations of it?" said his father.

"I don't suppose anybody knew what it was except me," said Martin. "I am sorry if you think I should not have played it. But really I had no time to think. I was nearly late, and on the organ there was only a book of dreadful extracts, chiefly by organists. But I will play something definitely sacred at the eleven o'clock service. That is if you would like me to play again."

"Thank you, dear lad, thank you. Ah, what a lovely morning! Look at the hills. 'I will lift up mine eyes to the hills.' How wonderful the appreciation of natural beauty in the Psalms is—'Sweeter also than honey'—so many of David's similes are drawn from ordinary, everyday sensations, but lifted up, ennobled, dedicated. But how was it you were nearly late? I looked into your room before I started for church and found you had already gone!"

"I went down to bathe," said Martin; "in fact it was

only the bell beginning that reminded me there was a service at eight."

Mr. Challoner looked at him a moment with a sort of appeal.

"But, dear Martin," he said, "you did not come without preparation?"

"I am afraid I did," said Martin, and the joy of his waking hours dropped utterly dead, while the hopelessness of the compact of the evening before rose close in front of him.

They took a turn up and down the lawn before going in, and his father very gently but very firmly impressed on him the positive sin of his omission. His voice trembled with the earnestness of his feeling, for to him the danger of coming to the Communion unprepared was as vital as the need for coming. He hated to say what he felt he must say: it was so soon after their compact to try to understand one another, to get on without perpetual correction and admonishment. But this could not be left unsaid. Once it occurred to Martin to tell him the truth, to say, "I came in order to please you; otherwise I should not," but the impulse passed. There was no need to give his father such pain as that: and he merely assented dully where assent was needed, said "Yes, I see" at intervals, and gave the promise required. But it was a dreary beginning to the day.

The Chartries pew, the only family pew remaining in the church, was well attended at the eleven o'clock service, Lady Sunningdale being, as usual, the brightest object present: indeed, among the rest of the congregation she resembled a bird of paradise which had by mistake found its way into a colony of sparrows. But what this violation of her habits in appearing so long before lunch had cost her none but her maid knew. However, there she was, and the colours of the spectroscope blossomed together in her hat, and in a fit of absence of mind, to which she was prone, she as nearly as possible put up a pink sunshade, forgetting where she was, to shield her from the sun, which was shining through a mauve-coloured saint onto the middle of her face in a manner which she felt to be aggressive and probably unbecoming. So she moved to behind the shadow of a neighbouring pillar, from where, looking at the organ, she could see who sat there.

"Too heavenly," she said in a shrill whisper to Stella Plympton. "Martin is at the organ. I'm afraid he won't play the Brahms, though. What a pity it is not Good Friday! he would be sure to give us the 'Charfreitag' music."

That, however, was not to be, and instead the familiar strains of "O Rest in the Lord" were the prelude, to which six choir-boys, four choir-men, including the carpenter, who in a fluty falsetto sang a steady third below the trebles and believed the result to be alto, advanced to their places. But Martin, in Lady Sunningdale's opinion, could do no wrong, and again she whispered shrilly to Stella:

"Is he not wonderful? That tune is exactly like the stained glass. It is absolutely the 'air' of the place. Look, there is Helen Challoner sitting with the choir. Is she not a dream? Tell Frank to look at her."

But this was unnecessary, as Frank Yorkshire was already looking. He was a rather stout, very pleasant-faced young man of about thirty, with smooth flaxen hair, rather prominent blue eyes, and an expression of extraordinary amiability, which his character fully endorsed. He was remarkably adaptable, and, while he would willingly talk flippancies with Lady Sunningdale, his tenantry adored him for his friendliness and his great common-sense if the baby was ill or the pig would not put on flesh. In other respects he was a baron of the realm, immensely wealthy and unmarried, so that he was perpetually drenched by showers of eligible girls whom aspiring mothers hurled at his head. These he returned with thanks uninjured.

He had, in fact, many pleasant qualities and one notable one, which Lady Sunningdale had already mentioned as being characteristic of him, namely, his undeviating pursuit of the first-rate. It was this which turned a character that would otherwise have been rather materialistic into something of an idealist, and supplied a sort of religion to a mind which otherwise, an extremely rare phenomenon, was completely atheistic, not with an atheism into which he had drifted from carelessness or *insouciance,* but with one that sprang from a reasoned and clear conviction that there could not possibly be any God whatever. On all other matters he had an open mind, and was extremely willing to adopt any opinion that seemed to him reasonable, but on this one point he was hopelessly bigoted. This reasonableness and willingness to be convinced had led people to suppose that he was weak. But this was not in the least true; he was only fair. Another quality, and a fine one, was his also: he was practically unacquainted with fear, either physical or moral, and would, had he lived in those uncongenial times, have gone as cheerfully to the stake for his entire absence of religious beliefs as he would now blandly uphold his abhorrence of sport on the ground of cruelty to animals in a roomful of hunting men. His faculty of reverence finally, of which he possessed a considerable measure, he exercised entirely over the talents of other people, on whatever line they ran. He knelt, for instance, at the shrine of Lady Sunningdale's acute perceptions, he hung up votive offerings to Martin's music, he even at this moment bowed the knee before the village carpenter, whose talent for singing the wrong note was of that instinctive and unerring quality which approaches genius.

He was a great friend of Martin's—Helen he only knew slightly—and after service desultory conversation in the churchyard ended in the twins going back to lunch at Chartries. Though Mr. Challoner was opposed on principle to anything however remotely connected with festivity taking place on Sunday, he raised no objection, merely reminding Helen that her Sunday-school class met at three. Lord Yorkshire, strolling by her, thought he heard a *nuance* of impatience in her assent, and his question had a touch of insincerity about it.

"Don't you find that charming?" he asked. "I think there can be nothing so interesting as helping to form a child's mind. It is so plastic—like modelling clay. You can mould it into any shape you choose!"

Helen glanced quickly at him.

"Do you really want to know if I find it charming?" she asked.

"Immensely."

"I detest it. I don't think they have any minds to mould. Why should one think they have? But they have shiny faces, and they fidget. And I point out Ur of the Chaldees on the map."

He laughed.

"I suppose the chances are in favour of their not having minds as you say," he remarked. "But I had to allow for your delighting in it, when I started the subject,

What do they think about then? Do they just chew their way through life like cows? You know some people don't chew enough. I expect Martin doesn't. But that is why he is so extraordinary."

There was no intention in this, and it succeeded. Anyone who admired Martin had found a short cut to his sister's favour.

"Ah, Martin never chews," she said. "I don't think he ever thinks: he just—just blazes. Now do tell me, Lord Yorkshire, because you know him well. He isn't stupid, is he, because he can't or doesn't pass examinations?"

"He couldn't conceivably be stupid, any more than I could be a Red Indian. But it is by a misguided ingenuity that he contrives not to pass examinations. It is hardly worth while doing it."

"Ah, do tell him that," said Helen. "I think you have influence with him."

"What on earth makes you think that?"

"He quotes you."

"Are you sure you do not mean he mimics me? He does it to my face too, so why not behind my back? It is quite admirable. Ah, I see he has shown you a specimen. Don't I talk wonderfully like him? But influence! —one might as well sit down and think how to influence a flash of lightning."

Helen considered this a moment.

"Well, there are such things as lightning-conductors," she said. "Besides, there are times when Martin isn't the least like a flash of lightning. He is often like a stagnant pool."

"I don't recognise that," said Frank.

"No, you probably have never seen it."

They had passed out of the narrow path from the

churchyard during this, and, their way lying across the open fields, Lady Sunningdale, as her habit was, annexed Frank as well as Martin.

"Dear Helen, it is too bad," she said as she manœuvred. "You will have to go back immediately after lunch. What is a Sunday-school? It sounds so beautiful, like a hymn tune. Yes, I adore church music; really there is nothing like it. And it was so wonderful of you to play the lucubrations of Mendelssohn, Martin."

"Yes, I felt that too," said Frank, in his low slow voice. "There was a stained-glass window just opposite me, which was exactly like the tone-colour of Mendelssohn. A figure which I take to have been a prophet, probably minor, in jewelled slippers, was directing an enamoured gaze towards a pink town—which may or may not have been the New Jerusalem. I always wonder where artists in stained glass get their botany from. Nameless herbs enveloped the feet of the minor prophet."

Martin laughed.

"I know that window," he said. "When I was little it used to come into my nightmares. Now it has become a daymare. I don't know which is worst."

Lady Sunningdale sighed.

"Church is very fatiguing," she said. "I had quite forgotten how tiring it was. I shall not go any more for a year or two. Dear me, these tiresome shoes! And my darlings wanted to come with me, but that isn't allowed, is it? It is only in Scotland that dogs go to church, I think. I went to Scotland once. I can't bear the Scotch. They are so plain and so extremely truthful. There is nothing in the least unexpected about them. Dear me, there's the other shoe. Yes, thank you, Martin. And they use a silly slang instead of talking English. Martin,

I had a talk to your father yesterday about you. I really think I made an impression."

"Telling the truth produces a very marked type of face," said Frank, "and in later life mutton-chop whiskers. That is why one always engages butlers with mutton-chop whiskers. They are sure to be reliable. Truth-telling is quite incurable, and so has a certain claim to distinction."

Martin listened to this with something of the air of a parrot "taking notice," and then turned to Lady Sunningdale.

"Do you really mean that?" he asked, eagerly.

"Yes, of course I do. It seemed news to him that playing the piano could be taken seriously. And he took me seriously. There are my treasures come to meet me. I am so hungry. Don't jump up, Suez Canal. My darlings!"

CHAPTER IV.

HELEN, as Lady Sunningdale had mentioned, had to start back again for her Sunday-school soon after lunch. They had all moved out under the cedar on the lawn, and when she rose, Lord Yorkshire also got up and offered himself as an escort. This was perfectly agreeable to the girl, though she wondered exactly how high Aunt Clara's eyebrows would rise if she knew that her niece might have been found walking on Sunday afternoon with a young man, who could not possibly be brought under the elastic bonds of cousinship. But the eyebrows of Lady Sunningdale, who, it must be supposed, was chaperone, remained low and level, and the two started.

Frank had been admirably entertaining in his own way during lunch, capping the extravagances of Lady Sunningdale with incongruities that rivalled her own, and giving wings of epigram and paradox to his speech; but Helen had received a very distinct impression that under his flippancy, which Martin imitated so faithfully, there lay something of sterling and very human solidity. And this unknown factor interested her quite apart from and much more than his conversational fireworks, which were as obviously superficial to the essential "he" as his eyebrows or moustache. Perhaps he also knew the unimportance of these leadings, for certainly as soon as they were alone, such coruscations died slowly down, and it seemed to Helen that a very pleasant mellow light, restful after fireworks, took their place.

"I think it is unkind of you not to admit me into the school itself," he was saying. "Why am I to be debarred from the knowledge of Ur of the Chaldees? Geography has an enormous fascination for me. I can pore for hours over maps of countries which I have never seen and almost certainly shall never see, just reading the names of unheard of places with gusto."

"Ah, you feel that too," she said. "Martin always tells me I am a gipsy. Certainly I want to wander, to go on just for the sake of going on. The exploration, that is the point. And I think it is the playing at exploration that is so fascinating in a map. Dictionaries too, new words. And best of all new books, with new ideas."

"There is one thing better," said he, "I cap your new books with new people, new ideas."

The personal note entered, however slightly into this, and Helen was silent a moment.

"Ah, but new books implies new people," she said. "Nothing can be more real than the people in some books."

"Quite true; and nothing can be less real than some people in real life. Do you know what I mean? One wonders with some people if there is anybody there. My impression is that there often isn't."

"I have an aunt——" Helen began and stopped, feeling that it was not quite kind to lay Aunt Clara on the dissecting table.

Frank guessed this.

"Ah, I have three," he said, "perhaps mine will do."

Helen laughed, and after a moment he went on.

"I believe that curiosity, which is a convenient expression to sum up all this passion for the new," he said, "is quite modern. I don't think at least that the generation to which our aunts belong had it, with certain adorable exceptions, like Lady Sunningdale, anything like to the extent we have it. What was good enough for our grandfathers was nearly good enough for our fathers. But what was good enough for our fathers is not nearly good enough for us."

She turned a quick luminous glance at him. He was talking about things that very much concerned her.

"Ah, that is interesting," she said eagerly. "Give me more news of that."

"It has struck you too?" he asked.

"Your saying it reminds me that I knew it all the time."

"I know what you mean. Yes, I think it is the case. At any rate, take yourself, Martin and me, all, I expect, quite normal people. Well, we all want to wander, to experience everything. We are probably not really afraid of any experience that could conceivably happen to us. And we claim the right to all experience. We claim the right to our own individuality too. It seems to us quite certainly ours, the only possession we have which is inalienable. We may lose everything else from our character to our teeth. But not our individuality. Do you remember how Magda throws her arms wide and cries 'Son' Io!'—'I am I!' That somewhat important point had never struck her father or mother. Poor things! They thought she was a sort of them. Is that bad grammar?"

Their way lay at this point through one of the game covers, and a sudden piteous crying, dreadfully human, arose from the bushes near the path. Helen stopped, with fright and horror in her face.

"A child! is it a child?" she asked.

"No, nearly as bad though, a hare," said he, and pushed his way through tangled bracken and brambles in the direction of the sound. In a moment he called to her.

"Will you come here, Miss Challoner?" he said. "Come round to the right; it is a clearer path."

She followed his directions, and found him kneeling a few yards off, holding in both hands a hare that was caught by the hind-leg in a horrible jagged-tooth trap.

"Pull the two sides of the trap apart," he said, "as quickly as you can. Be quick. The poor brute is struggling so, I can hardly hold it."

His voice was so changed that she would hardly have recognised it. It was no longer low and courteous, but sharp and angry. She knelt down by him, and exerting her full strength did as he bade her. The leg was caught only by the skin, and holding the animal in one hand he gently disimpaled it, where the iron teeth had clutched it.

But just as it was free a sudden tremor of nerves passed through Helen at this humane surgery; the trap slipped from her hand, and caught Frank's finger just at the base of the nail. He took his breath quickly with the pain, and let go of the hare, which, none the worse, ran off up the winding path down which they had come.

"I must trouble you to open the trap once more," he said, the blood streaming from his finger. But now his voice was quite normal again.

"Oh, I'm an absolute fool!" cried Helen. "Oh, I'm so sorry;" and again she wrenched the trap open.

Frank was rather pale, but he laughed quite naturally.

"Thank you so much," he said, as she released his finger. "What strong hands you have! But I should dearly like to clap that thing on the nose of the brute who set it. What an infernal contrivance! How can men be such butchers? I shall take it and show it to your uncle."

He shook the blood off his finger and bound it tightly round with the handkerchief.

"Oh, Lord Yorkshire, I'm so sorry," said Helen again. "I am an absolute born idiot. How could I be such a fool?"

He laughed again.

"My dear Miss Challoner," he said, "nothing whatever has happened which can justify your violent language. Besides, it would have been worth while to set that poor jolly beast free at the cost of real pain, and not just a finger-scratch. Well, we've vindicated the liberty of one individual anyhow. Did you see its eyes? They said, 'I am I,' like Magda."

He held the bushes back for her to regain the path.

"But you'll have your finger attended to?" she said.

"Yes, at once, please. I'll ask you to tie it rather tighter, if you don't mind the sight of blood. I always think blood is such a beautiful colour," he chattered on, to prevent her apologising further. "One talks of a blood-red sunset and admires it, and dragon's blood china, but when it comes to the real article so many people shrink from it. That's better, thanks; that's excellent. I assure you it is nothing at all."

His manner was so entirely natural that there was nothing left for her except to be natural too, and they walked on out of the cool green-shadowed path, flecked here and there with the sunshine that filtered through the trees that met above them into the blaze and brightness of the fields that bordered the churchyard.

"Yes, the cry of Magda for her right to her own individuality," he said. "At last this generation has said, 'I will lead my own life, not the life dictated to me by other people.' I wonder what we shall make of it."

Helen looked at him again eagerly.

"And do you mean that the assertion of one's own individuality is a duty?" she asked.

"Ah, that is a difficult question. Certainly I think there are—are indications that one is supposed to play one's hand for all it's worth. But duty? Probably you and I mean different things by it."

"I mean the will of God for me," she said, simply.

They paused at the gate into the churchyard, and their eyes met. It seemed to Frank that she waited for his answer with some eagerness. And he shook his head.

"No, I don't mean that," he said.

She held out her hand to him.

"I'm sorry," she said.

"So am I, very sorry indeed. But I can't help it."

Her eyes wandered over the woods behind him, then came back to his face.

"No, I recognise that," she said. "Good-bye, Lord Yorkshire. Thank you so much for coming with me. And please have your finger attended to."

She smiled at him, and went up to the churchyard path towards the shining corrugated iron room. As she passed the walk leading to the Vicarage she met her father.

"You are nearly ten minutes late, Helen," he said.

"I know, dear, I am sorry. But you know you are late, too."

He did not smile.

"I was detained by other parish work," he said. "I was not amusing myself. Pray do not delay any longer."

The evening meal on Sunday evening at the Vicarage was of a strictly Sabbatical order, and consisted of cold things to eat, and no waiting on the part of servants. It took place late after evening church, and had to Martin's mind a dreariness of its own, an individuality (to which Frank would have said it undoubtedly had a right) which marked it off from all other meals. Everyone was fatigued with the exercises of the day, and though they were religious exercises which had produced that fatigue, they brought with them a tendency which made cheeriness difficult. However, cheeriness was not a quality exactly encouraged by Mr. Challoner on Sunday, so perhaps that was all for the good. But this evening, Martin, who had spent the whole afternoon at his uncle's, coming back only just before supper, was conscious of a Sunday easily got through, and was chattering on with a good deal of rather thoughtless enjoyment about Lady Sunningdale, every now and then mimicking with extreme fidelity some more than usually incoherent speech of hers, in which Wagner, her dogs, South Italy, her husband, egg-shell china and scandal were about equal ingredients, without noticing a somewhat ominous gravity that was deepening on his father's face.

At length Mr. Challoner spoke, interrupting him.

"There, dear Martin, is not that enough? It is Sunday evening, remember. Cannot we find something rather more suitable to the day to talk about? And you would scarcely like Lady Sunningdale, who is so good to you, to know that you imitate her."

"Oh, she is always insisting that I should do it to her face," said Martin. "I often do. She shrieks."

"That is enough, I think, Martin," said his father again, mindful of their compact of the evening before, and determining to be gentle. "Have you only just come back?"

"Half an hour ago," said Martin, the gleam in his eye suddenly quenched, for he knew what the next question must be.

"Then you did not go to church this evening?" asked his father.

"No, I had been twice."

Now, Mr. Challoner had been from church to Sunday-school, and from Sunday-school to church practically since eight that morning, and it was not in the least unreasonable that he should be tired with so many busy hours in ill-ventilated places on so hot a day. The effect of this tiredness on him, as on most of us, was shown in a tendency to that which when it occurs in children their elders label "crossness." And he answered in a tone in which that very common emotion was apparent.

"I was not asking you to justify your absence," he said; and the meal proceeded in rather dreary silence.

Then two small incidents happened. Martin dropped a plate with a hideous clatter, and a moment afterwards upset a wineglass which he had just filled with claret all over the table. He apologised and wiped it up, but unfortunately, looking up, he saw his father's face wearing such an extraordinary expression of true Christian patience that for the life of him he could not help giving a sudden giggle of laughter. He could not possibly have helped it; if he was going to be hung for it he must have laughed.

Now the laughter of other people, when we ourselves do not see anything whatever in the situation to provoke mirth, is one of the authentic trials of life, especially if one half suspects, as Mr. Challoner did now, that one is, in some manner inexplicable, oneself the cause of it. It was therefore highly to his credit, that remembering the interview he had with Martin the night before, he could manage to keep inside his lips the words that tingled on his tongue. Of more than that he was incapable; he could not just then be genial, or start a subject of conversation; he could only just be silent.

Martin could easily manage that; his last observation had not found favour, and he held his tongue and ate large quantities of cold beef. Helen, sitting opposite her father, in the absence of Aunt Clara, who was spending the Sunday away, had also nothing apparently which she considered as suitable, and the meal proceeded in silence. Then after a pause of considerable length, she raised her eyes, which so happened to catch Martin's, who was still struggling with his unseemly mirth. At this moment also her father looked up and saw a glance which he interpreted into a glance of meaning pass between them, a thing irritating to the most placid temperament. He saw too the corners of Martin's mouth twitching. This was too much.

"I will not have that sort of thing, children," he said, his voice rising sharply. "It is an exceedingly rude and vulgar thing to exchange glances like that."

Martin's merriment was struck as dead as beech-leaves in frost.

"I was doing nothing of the kind," he said, his temper flashing out. "Helen looked up at the same moment as I looked up. We all three looked up, in fact; it was purely accidental."

Helen was vexed that Martin should speak so, but felt bound to endorse him.

"Indeed, father, it is so," she said.

Again the silence descended; and Martin seeing that both his father and sister had finished their meat, changed their plates and arranged the second course. After a very long pause their father spoke again.

"I should have thought my children might have had something to say to me in the evening when they have left me alone all day, enjoying themselves elsewhere. Has nothing happened to you since breakfast which I am worthy of hearing?"

Martin's intolerance of this injustice again stung him into ill-advised speech.

"I tried to tell you what I have been doing," he said, "but you stopped me. You said it was unsuitable;" and his handsome face flushed angrily.

Then a thing unprecedented happened.

"I beg your pardon, dear Martin," said his father.

Helen was engaged next morning in the fragrant labour of picking sweet peas when a maid came out of the house to say that Lord Yorkshire was there. Her father and Martin she knew were both out, and she went in to see him, concealing from herself the quite perceptible thrill of pleasure that the announcement had given her. She was, as usual, hatless, and her hair was in golden disarray from the breeze, and as she went towards the house she took off her gardening gloves, trying by sundry pats and pokes to give it some semblance of order. She was not very successful in this, nor need she have been, for she looked to him like some beautiful wild flower when she entered.

"I ought to apologise for coming at this unearthly hour," he said, "for my only excuse is that Martin left a book of music at Chartries, and having an idle morning I thought I would bring it over."

Helen was delighted to see him, and since it would have been ungracious to convey the impression that this morning visit was a bore, especially since it was not, she took the straightforward line.

"How good of you!" she said. "And the finger?"

He held up a bandaged hand.

"I am only reminded of it by that," he said.

"I am so glad. Isn't it extraordinary that anyone could be so awkward as I was? I am always dropping and spilling things. Martin used to say, 'It is a lovely day, let us go and spill something.' But he is much worse than I am, really. Do come and look at the garden. It is really pretty."

"And are you gardening?" he asked, glancing at the gloves.

"Mildly. I am really only picking sweet peas. It is so nice of them—the more you pick the more they flower."

She picked up her basket as they walked out, and held it up to him.

"How energetic of them!" he said. "Ah, what a delicious smell! That reminds me of lots of nice things; it will now remind me of one nice thing the more. Smell is the keenest of all the senses to remind one of things. Sight and hearing are not nearly so intimate. And Martin is out?"

"Yes, he went to try and get a fish. But there is too much sun."

"I am delighted to hear it," said Frank.

"I think I am too, really," she said. "But I do like the dear boy to be pleased."

"Well, I hope we are all going to please him," said he. "For the combined armies are going to advance and rescue him. Lord Flintshire, Lady Sunningdale, and, in my own humble manner, myself, are all going to try to get your father to allow him to study music in earnest. In fact, I am a sort of skirmisher in advance of the heavy —of the main body. It is my business to bring on the general engagement by asking him to stay with me in London, and bringing some people who really know to hear him play."

Helen turned a radiant face on him.

"Ah, that is good of you," she said; "and it is really angelic of me to feel that, as I shall be left here all alone."

"But the scheme includes you. Lady Sunningdale is writing to you to ask you to come up with him and stay with her for a week or two. I hope you will say 'yes.'"

Helen gave a long sigh, as Moses perhaps sighed on Pisgah.

"I don't know if I could manage it," she said, "though it would be heavenly. Perhaps, as Aunt Clara comes back in a day or two, I could leave father. But I don't know. Oh, I should enjoy it!" she cried.

"I expect you have a very fine faculty for enjoyment," said he.

Again the personal note entered, but this time it did not make her pause.

"I? I should just think I had. And I love London in little raids like this, it is so full of charming things to do. But Martin—it is good of you, Lord Yorkshire. And do be very good for him. Do use your influence with him. Do make him at any rate work hard to pass his examination at Cambridge first. It would make everything so much easier, so much happier."

"For him?" he asked, with a marked intonation.

"Yes, and for all of us."

He looked at her gravely.

"That sounds worth while," he said.

He let that string vibrate, as it were, for a moment or two, and then passed on.

"But what becomes of the liberty of the individual which we talked of yesterday?" he said. "To influence anybody always seems to me a slight infringement of rights. One imposes one's personality—such as it is—on another."

"Ah, but in a good cause, to show him the stupidity of not passing examinations. Surely that is a rule absolutely without exception, that it is always wise not to be stupid."

He laughed: Helen with her direct vivid personality, seemed to him unlike anybody else he had ever seen, with the exception perhaps of her twin. The extraordinary and rather rare charm also of perfect naturalness, not the assumption of it, was hers also.

"Well, it is certainly hard to think of any exception to that rule," he said, "though one always distrusts rules without exceptions. It seems so very unlikely that they should exist, considering how utterly different every one person is from every other. On the face of it it seems impossible."

This had aroused another train of thought in the girl.

"Oh, nothing would be impossible, if one were wise," she said. "Oh, I hate fools. And I am one."

And she snipped viciously among the sweet peas.

He followed this with some success.

"Was the Sunday-school very stupid?" he asked, sympathetically.

"Hideously—quite hideously. How clever of you to guess! It was also extremely ugly. I don't know which I dislike most, ugliness or stupidity. In fact, they are difficult to tell apart. Yet, after all, beauty is only skin deep."

"But what has that to do with the wonder of it?" he asked. "That particular proverb seems to me about the silliest. Why, the most subtle brain in the world is only a few inches deep, and as far as measurement goes, it is about the same depth as the most stupid. Or would you say that the beauty of some wonderful evening-moment of a Corot was only skin deep, the depth of the paint on the canvas? Surely not. It has all the depth of beauty of the summer night. No, that proverb is perfectly meaningless, and was probably invented by somebody more than usually plain."

Helen's basket of sweet peas was full, and she emerged from the fragrant tangle of the garden-beds and strolled with him up the lawn, her face on flame with what he had called curiosity. That divine moment when a girl becomes a woman, when all she has drunk in all her life begins to make products of its own, had just come to her. And at this psychological moment he had come too.

"But surely one sees very beautiful people who are very dull, very stupid, very wicked even," she said. "Is not that what the proverb means, perhaps, that as far as beauty itself goes, it is only a very superficial gift?"

He shook his head.

"Look at that splendid Gloire de Dijon," he said. "It may be very stupid, very dull, very wicked, so far as we know. But that does not concern us. It is beautiful, and its beauty does not anyhow touch us only superficially, but very deeply. Does not beauty stir in you some chord of wider vibration than any purely intellectual quality? Some—how shall I say it?—some longing for the infinite?"

Again their talk had taken the bit in its teeth, and as she gently fingered the rose he had pointed to, her lips drew themselves into a quivering curve of extraordinary tenderness.

"Ah, yes, yes," she said. "I could kneel down and thank God for it."

He looked at her gravely, remembering the conclusion of their walk the afternoon before.

"You are very much to be envied," he said. "With my whole heart I congratulate you."

She raised her head, dismissing the gravity of the last minute.

"Ah, but the Sunday-school," she said.

"But I envy that too," said he. "It as well as you has its *beaux jours*. You would not grudge it them."

She laughed.

"Ah, you have committed an inanity," she said. "I was so afraid you were a person who never said anything stupid. But to pay compliments is stupid. And now I have been rude. That is even more stupid."

"I think it is," said he, "because it is also unnecessary."

There was a further challenge in this, but she did not take the glove he had flung, and having reached the tree at the end of the lawn, underneath which three days ago the ill-fated "Mill on the Floss" had lain, they turned back again towards the house, and she directed their talk, like their steps, in another direction.

"It is good of you, I mean about Martin," she said. "That is just what he wants, to go among people who will take him and his music seriously, not gasp because he plays extremely fast. No one here really knows the difference between 'Rule Britannia' and the 'Dead March.' And yesterday—oh dear, oh dear!"—and she broke out laughing.

"There isn't much," observed Frank parenthetically. "But please tell me about yesterday."

"I think I must, because, though you will laugh, you will laugh kindly. It was at the early service, and the dear boy played the overture to 'Lohengrin' as a voluntary, and my father thought it wasn't quite suitable."

He considered this a moment.

"Do you know I don't think I want to laugh at all?" he said. "I understand perfectly."

"But Martin didn't. That was so funny."

"No, he wouldn't. That is one of the penalties of genius. In fact it is what genius means. It is having one point of view so vivid that all others are dark, invisible beside it. And genius is always intolerant."

Her eye brightened.

"I don't know if you know or not," she said, "but I expect you do. Is Martin really all that—dear, stupid old Martin?"

"I believe so. We are going to get him to London to find out. You will give him my message, won't you? I go up to town to-day, and he may come any day he likes, the sooner the better. Lady Sunningdale is writing to you."

"Oh, it would be heavenly!" said she.

He took his leave soon after, and went back to Chartries for an early lunch, since Lady Sunningdale, who never started anywhere in the morning, unless it was impossible to get there otherwise, had retained his services in order to minimise the dangers and difficulties incident to travel by rail with Suez Canal and Sahara. For Sahara had an unreasoning dislike of locomotive engines, which had never at present hurt her, and always tried to bite them; while Suez Canal, whenever it was feasible, jumped down between the platform and the train, and smelled about for whatever there might be of interest among the wheels of the carriages. In addition to these excitements their mistress never moved without a tea-basket, a collapsible card-table, which usually collapsed, a small library of light literature and a jewel case, so that the tedium of a journey in her company was reduced to a minimum, since, when the train was in motion, these recreations could be indulged in, and when it stopped there was more than enough to be done in collecting these priceless impedimenta to prevent any companion of hers from feeling a moment's boredom that arose from idleness. She also could hardly ever produce either her own or the dogs' railway tickets when called upon to do so, thus giving rise to games of hide-and-seek all over the carriage.

And to-day in addition Frank had something very considerable of his own to think about—something that made him very alert, yet very inattentive—that brightened his eye, yet prevented him seeing anything. And he could almost swear that the odour of sweet-peas pervaded the railway carriage.

Martin meantime was spending the morning on the banks of the stream which had given him those good moments early the day before. But to-day the sun was very hot and bright, and after an hour's fruitless but patient attempt on the subaqueous lives, he abandoned the vain activity of the arm, and, with the vague intention of returning home and getting through some Æschylus before fishing again towards evening, sat down to smoke a cigarette in the fictitious coolness bred by the sound of running water, preparatory to trudging back across the baked fields. Tall grasses, mixed with meadow- sweet and ragged robin, moved gently in the little breeze that stirred languidly in the air, but the sky was utterly bare of clouds, and stretched, a translucent dome of sapphire, from the low-lying horizon of the water-meadows on the one hand, up to the high yellowing line of the downs on the other. At his feet flowed the beautiful stream, twining ropes of shifting crystal as it hurried on its stainless journey over beds of topaz-coloured gravel or chalk that gleamed with the lustre of pearls beneath the surface. Strands and patches of weed waved in the suck of the water, struck by the sun into tawny brightness, shot here and there with incredible emerald, and tall, brown-flowering rushes twitched and nodded in the stress of the current. Suspended larks carolled invisible against the brightness of the sky, swallows skimmed and swooped, and soon a moorhen, rendered bold by Martin's immobility, half-splashed, half-swam across the stream just in front of him. And he thought no more of the fish he had not caught, but sat with hands clasped round his knees, and, without knowing it, drank deep of the ineffable beauty that was poured out around him on meadow and stream and sky. Every detail, too, was as exquisite as the whole: the yellow flags that stood ankle-deep in the edge of the river were each a miracle of design: the blue butterflies that hovered and poised on the meadow-sweet were more gorgeous with the azure of their wings and white and black border than a casket of lapis-lazuli set with silver and shod with ebony.

By degrees, as he sat there, his cigarette smoked out, but with no thought of moving or of Æschylus, the vague and fluid currents of his mind that for years had coursed through his consciousness, though he himself had scarcely been conscious of them, began for the first time to crystallise into something illuminating and definite. Like some super-saturated solution of chemical experiment his mind, long crying out for and demanding beauty, needed but one more grain of desire to render its creed solid, and to himself now for the first time came the revelation of himself, and, like a spectator at some enthralling drama, he watched himself, learning what he was, without comment either of applause or disgust, but merely fascinated by the fact of this new possession, his own individuality, and, even as Frank had said to Helen only yesterday, his own inalienable right to it. It was none other's but his alone: there was nothing in the world the same as it, since every human being is a unique specimen, and, bad or good, it was his own clay, his own material, out of which his will, like some sculptor's tool, should fashion a figure of some kind. And everything he saw, the yellow iris, the blue butterfly, the water-weeds, were in their kind perfect: their natural growth, unstunted by restraint or attempt to control them into something else, had brought them to that perfection; and was it conceivable in any thinkable scheme of things, that man, the highest and infinitely most marvellous work of nature, should not be capable of rising, individual by individual, to some corresponding perfection? Soil, sun, environment were necessary: the flags would not grow in the desert, the lark would not soar nor carol in captivity; but given the freedom, the care or the cultivation which each required, every living and growing thing had within itself the perfection possible to itself.

Up to this point his thought had been as intangible as a rainbow, though, like a rainbow, of definite shape and luminous colour, and showed itself only in a brightened unseeing eye and in fingers that twitched and clutched till the nails were white with pressure round his flannelled knee. Then suddenly the crystallisation came, ungrammatical but convincing.

"It is me!" he said aloud, as Magda had said it.

In a moment the whole solution was solid.

Beauty. That was the food for which every fibre of his nature hungered, and with which it would never be satiated. Long ago he had known it, but known it second-hand; known it as in a dream, when he quoted Browning three days ago to his sister. But that dream, that second-hand information, had become real and authentic. No matter how trivial might be the experience, that was what he demanded of all experience—whether he ate or drank, it was beauty he craved; whether he ran or sat down he knew now that in so far as it was consciously done, it was the thrill of speed, the content of rest that he demanded of the function. Then suddenly he asked himself what he demanded in the exercise of the highest function of all, that of worship—was it the pitch-pine pew, the magenta saint, the tuneless chant? Was it the fear of hell, the joy of an uncomprehended heaven, even though the gate-stones of the New Jerusalem were of jasper and agate? Not so; for what did he worship? Absolute beauty, that quality of which everything that is beautiful has some grain or mirrored reflection. That was God, the supreme, the omnipotent, present in all that was beautiful, just as much as He was present in the breaking of the Bread, and the outpouring of the mystic Wine; for all was part of Him.

CHAPTER V.

THE big drawing-room at Yorkshire House was full to overflowing, and for the avoidance of asphyxiation the six long windows that looked onto the Green Park were all open. Louis XVI. candlesticks, converted to the more modern use of electric light, were brilliant on the crimson satin of the walls, and a couple of dozen rows of chairs, all occupied, were directed towards the end of the room where the Steinway grand stood. Behind the chairs there was a throng of standing folk; but, except for the voice of the piano, no sound broke the stillness. A quarter of an hour ago the smaller drawing-room opening out of this had been full of chattering groups, but now it was completely empty except for some half-dozen people who had been unable to find a standing-place in the larger room, and crowded as near as they could to the doorway. But the last human voice had been that of Martin.

"I'll play it if you like," he said, "but it will take nearly half-an-hour."

Then he sat down, and, since he had played before, a hush most abnormal during the ordinary piano solo fell on the "party" which had been invited in after dinner. Many, no doubt, were unmusical, but more, since it was Frank's house and it was he who had invited the guests, had some instinct for perfection, that bond that joins together all artists. Lady Sunningdale of course was there, and had early established herself in a front row; and Helen, who was under her chaperonage, sat next her. At the end of the fourth *étude* of Chopin's, she had said to Martin:

"Martin, play the Brahms' variations," and the demand had led to his word of warning. But warning was not needed—if the piece was going to take an hour, no one would have complained.

Frank, knowing the acoustic properties of the room better than Lady Sunningdale, had placed himself in the seat of the second window, with Karl Rusoff beside him. He had himself not felt the slightest hesitation in asking the great pianist to listen to the recital of this wonderful *débutant,* and Karl's absolute silence at the end of the variations convinced him that he had been right. And as the last glorious fantasy vibrated and died on the air, while the crowd burst gloves in applause, he turned to him.

"Well?" he said.

Karl Rusoff nodded his great grey head up and down.

"Ah, my dear young friend," he said, "I usually think it very clever to unearth a genius. But with your genius it needed no cleverness. Shall I tell you what will happen? We—the pianists, I mean—with our nimble professional fingers will in a year's time be fighting each other for seats at his concerts, if he is kind enough to give any. Let him give one, however, just to show us, to—yes, I mean it—to let us weep over our own deficiencies. Fire, my God, what fire! But I hope he won't give many: he ought—I only say he ought—to be too busy with his own work. As regards his piano playing, of course you were right. Who has taught him? Nobody, I tell you. How can you teach *that?* Will I teach him? Certainly I will, as Molière's housemaid taught her master. He does a hundred things quite wrong. But—ah! a big but!"

Martin had risen and bowed his thanks to the storm of applause, but his eye sought the corner where Karl Rusoff sat with his great grey leonine head, and his grey eyes gleaming through his spectacles. The latter rose, and came up the gangway between the chairs and the wall towards him, and shook hands with him.

"Mr. Challoner," he said, "that was a great treat to me. Thank you. You can play what is really difficult, magnificently. Now, my dear young man, I want to ask you a great favour. Attempt something much more difficult, that is to say something where the notes are quite easy, but where the rest, which is everything, must be a poem. Play, if you happen to know it—really know it, I mean—Chopin's 15th prelude, the rain on the roof."

Martin looked round the room, but nobody had moved from his seat, except Frank, who had followed Monsieur Rusoff.

"Yes, I know it," he said. "But are you sure you really want me to play again?" he asked, with the charming horror that a nice boy has of being a bore. "Are you sure they aren't sick of me?"

"No, do play again, Martin, if you will," said Frank, who had followed Karl. "We can really stand a little more."

"I have asked him to play the 15th prelude," said Rusoff.

"Ah, yes, do," said Frank.

So the rain beat, the gutters choked, the chariots of God thundered overhead, one ray of sunlight gleamed, and again the rain, pitiless and slow, spoke of an alien land. And at the end, in the moment's silence, more appreciative than any applause, which followed, Martin's glance again sought the great pianist, and with a sudden spasm of joy, so keen that for a moment he thought he must shout or laugh, he saw that Karl Rusoff had taken off his spectacles and was wiping his eyes.

The party that Frank had brought together that evening was very typical of his tastes and of the position which he held in the world. Though only thirty, thanks partly to his great wealth, which was always completely at the service of any artistic cause, but chiefly to his own exquisite and unerring artistic sense, he had now for some years been a sort of accredited godfather to any new talent, and for anyone to "come out" at his house was a guarantee that the aspirant was to be taken seriously. During the three months of London season he gave a succession of evening parties which all had some definite *raison d'être,* chiefly musical. And to-night he had taken special pains to get all the right people, with the result that there were not perhaps a dozen people in London, whose opinion was worth having, who were not there. And the opinion for once was practically unanimous; for though Claud Petman, plump and short-fingered, had something to say to Henry Runton about the lack of finality in the determination of his key-colour, and Henry Runton, over ortolans, agreed, with the additional criticism that his phrasing of the fourth variation was a little pulpy, yet the fact that they were critics rendered it obligatory on them to criticise. But they had but small opportunity to express these fine differences of opinion to Martin himself; for Lady Sunningdale, on the conclusion of the prelude, beckoned imperatively to her "monster," and made a brilliant group round him. She had taken it into her head that she had "discovered" Martin, and told everyone so.

"My dear, I assure you I gasped," she said to Karl Rusoff. "There he was in a poky little room, furnished entirely with prayer-books, in a dreadful parsonage, playing on a cracked tin-kettle of a piano, and playing as he played to-night. Then in the middle his father came in and said, 'Go and do your Hebrew-Greek, instead of wasting your time at the Jews'-harp.' Such a strange man, Flints' brother, you know, and lives, I believe, entirely on locusts and wild honey, and wears broadcloth, or is it sackcloth? Something very thick and imperishable anyhow. Such a beautiful life, but ascetic not artistic, Mendelssohn and pitch-pine, you know. Of course I saw at once how priceless Martin was, but we had the greatest difficulty in persuading his father to let him come up to London. He thinks all artists will go to hell, if they have not already gone there. Yes. I didn't bring my darlings to-night, because they always bark when anyone plays the piano, and Suez Canal is so shrill. But is not my monster too wonderful? And now I must go. I never get to bed till it is time to get up, and I shan't sleep one wink after the music. I never do. Where is Helen? Yes, she is Martin's twin. Why are not all twins like that? Supper. How nice! I am famishing. Music always takes so much out of one. Yes, pray take me in to supper, Monsieur Rusoff, and let us put it back. Martin, don't dare to leave my side for a single moment."

Frank, in the meantime, had found a chair next Helen. The girl looked divinely happy; her pride in Martin, her intense pleasure in the wonderful reception he had been given, flushed her cheek with excitement and sparkled in her eyes. Frank had not had an opportunity of speaking to her the whole evening; and now, as he was making his way towards her through the crowd, delayed every other moment by some acquaintance or friend, he met her eye long before he was within speaking distance, and as he smiled back in response to her, something suddenly thumped softly and largely on his heart, as if demanding admittance. At last he reached her, and she looked at him with her direct child-like gaze.

"Thank you," she said, "thank you most awfully."

He laughed, pretending not to know what she meant.

"Ah, we are all thanking Martin," he said, "and those who know best, I think, thank him most. Karl Rusoff, for instance."

"Then you were right," she asked. "There is no mistake? He is really of the best?"

"Yes, that is Monsieur Rusoff's opinion."

"I should like to kiss him," said Helen.

"Shall I fetch him?" asked Frank.

"Not this moment. Go on, Lord Yorkshire."

"That is a good deal already. And he will take him as a pupil, he says. He has not consented to take a pupil for years. Now we have to consult—how is that to be managed?"

Helen's face fell for a moment.

"It must be managed," she said. "I will write to father to-morrow, telling him all that has happened. You must write too; Lady Sunningdale must write. Poor father! we must give him no peace till he lets Martin study."

"You must think it over, and tell me if I can be of any use," said he. "I am entirely at your disposition. Anyhow, there is a fortnight for him in London. And you—you came up to-day, did you not? Ah, before I forget: Lady Sunningdale is coming to my box at the opera to-morrow night. Please come too. She, Martin, you, I. Just we four."

Those last three words gave him extraordinary pleasure.

"But are you sure you have room for me?" asked Helen. "Lady Sunningdale is so kind; she is dumping me at all her friends' houses, upsetting their dinner-tables right and left, and there is no earthly reason to suppose they want me!"

"I want you," said Frank simply; and again the words pleased him.

"Thank you very much. Where is she, by the way? Will you take me to her? She probably wants to go home. I see people are leaving."

"It is conceivable she is having supper," said Frank gravely. "Let us go and see."

Karl Rusoff attended to Lady Sunningdale's wants, which were rather extensive, but lingered after she had left, and when the rooms were growing empty he came up to Martin.

"My dear Mr. Challoner," he said, "I am sure you

have had enough compliments paid you by this time. So allow a very rude old Russian, who has no manners at all, to take you into a corner and talk to you for a little."

Martin turned a brilliant glance, vivid, and full of huge youthful enjoyment on him. He knew, he could not help knowing, how complete had been his success; and coming straight from the country and from that home where he was officially an idler, almost a black sheep, into this cultured critical world, the knowledge had somewhat intoxicated him. It was like coming out of some dark dripping tunnel into the light of a noon day, and flying along through a kingdom that was his. For he, he had been the central figure, round him had crowds collected, for him ears had been alert, and applause had burst. Artist as he was by nature, and caring therefore infinitely more for his art than for any adventitious success that he might achieve by it, he would not have been human, and certainly not young, if this evening had not been honey and wine to his boyish heart. For, except to the sour, success is sweet, and it is only the cynic and the unsuccessful who affect to find applause hollow. And Martin was emphatically neither cynic nor sour; the world seemed to him the most excellent habitation. But he detached himself at once from the group which was round him; he was still sufficiently master of himself to know that it was probably better worth his while to listen to Karl Rusoff talking sense than to anyone else who might have pleasant things to say, and they passed out of the supper-room into the now deserted room where he had played.

"Now, my dear Mr. Challoner, listen to me," said Karl. "Probably a hundred people this evening have told you that you are a very wonderful young man. That cannot help being a pleasant hearing, but——"

He looked at Martin's radiant face and paused.

"Ah, my dear boy," he said, "I will talk another time, I think. Go and listen to what everybody else has to say to you. Drink it all in; enjoy yourself. I am too serious; I can wait."

"But I would sooner listen to you," said Martin.

"Are you sure? Are you really sure?"

"Quite. Absolutely."

"Well then, in the sacred name of Art forget all the pleasant things that have been said to you. So many of these delightful people do not *know*. Our charming Lady Sunningdale even, she does not know. She appreciates, I grant you, but that is all."

Martin's face had grown quite serious; the brightness in it seemed to have ceased to be on the surface only; it glowed beneath like the core of a prospering fire.

"Tell me what to do then," he said.

"Work, and live also. Do not forget that any experience in life, so long only as it is not sensual, for whatever is sensual blurs and deadens the fineness of any gift, gives richness and breadth to your power in music. Live then; live to your utmost and your best. Do not be afraid of anything; neither the bitterest sorrow that the world holds, nor its most poignant joy, can bring you anything but good, so long as you embrace it willingly, passionately. But shun a sorrow or a joy, and you are clipped, maimed, blinded."

The old man spoke with extraordinary fire and emphasis, and the intense, eager gravity of Martin's face deepened. Here was a coherent code which, summed up, strung together his own musings by the river brink.

"Am I then to—am I to take all that comes," he asked,

"and trust that it will somehow make grist for my own little mill?"

"Ah, you understand," said Karl. "I see you have thought of it before. But never call your mill little. If it is little, you may be sure that others will label it for you. And if it is not little—then down on your knees and thank God. Ah, my dear boy, you are all that you are. Make the most of you. Assume there is something——"

He paused a moment.

"And I will endorse it," he said.

Again Martin looked at him with that lucid glance as transparent as running water.

"Yes, I will endorse it," he repeated. "And if anyone dishonours your cheque, I will pay it."

Martin gave a long sigh.

"You believe in me?" he asked, almost in a whisper. The rest of the triumph of the evening, the silence, the applause, were pale and dim to him as compared with this. The sun was rising on a dream that he had scarcely dreamed, and it was not a dream but a reality.

"I believe in your possibilities," said Karl. "I believe you can be—well, a musician. Now, as regards another point. I have been asked whether I will take you as a pupil. On my part I ask you to come to me. I have not taught for some years, but I rather suspect that one's power of teaching increases not by teaching but by learning. So I may be perhaps of some use. There are certain things I can tell you. Come and learn them. On the whole it is worth your while. Even for a poet the alphabet is necessary. He uses the same twenty-six letters as the Bradshaw."

Martin could not speak for a moment.

"Some day I will try to thank you," he said at length,

"But not by words. I don't think you want that, and also it would be idle for me to do it."

He paused again.

"But at present, you know, I am not even certain that I shall be allowed to study. I—I am very stupid, you know: I can't pass examinations, and my father is most awfully keen about them. In any case I expect I shall have to finish my time at Cambridge."

Rusoff rose. Absurd and almost criminal as this seemed to him, he had no right whatever to express that to Martin.

"Ah, then, go back to Cambridge like a good boy, and do whatever has to be done. Forget also almost everything that has occurred to-night. You have won a great deal of applause: well, that is very easy to win, and in itself it is worth absolutely nothing. In so far as it encourages you to good work, whether it is now or in the immediate future at Cambridge, or eventually in music, there is no harm in it; but the moment it breeds in you any slackness, or the feeling 'this will do for them,' it is a poison, an insidious narcotic poison."

He laid his hand on the boy's shoulder.

"It is not by applause," he said, "it is not by any help really that I or anyone else can give you, that you may become great. It is in yourself alone that the power lies; and it is by your life, by your industry, and by the fulness and completeness of your experience and your sympathy, that you will be able to get hold of that power. For your warning, I tell you that it is no easy task—that mining in yourself—you will have to think and struggle and despair before you can bring your own gold to the surface. You will also have to find your voice by patient, unremitting work. You cannot make others feel unless you feel yourself, and you have to learn how to feel: it is not so easy. Again having learned that, you cannot convey what you feel until you have learned speech. And for your encouragement I believe—or else I would not accept you, much less ask you to be my pupil—I believe that you will be able to do so. You have perception: you can interpret others as I have heard to-night. So that some day you may write that which will give tears or laughter to those not yet born. Good night."

The summer and the season were at their mid-most, but though the former had been fine, the latter at present had been rather objectless. Balls, concerts, parties, all the various devices by which the crowd believes it amuses itself, and without which it would certainly be bored, had occurred with their usual frequency; but up till now no bright particular star had arisen to draw the eyes and the thoughts of all to itself. There had, in fact, been no "rage," and neither book, play, violinist or traveller, nor even a cowboy from the remote West, had appeared to fill the invitation cards and usurp the thoughts of emigrant London. Why nobody had invented something by this time was not clear, for absolutely anything in the world can become the rage of one season, to be dropped either like a hot potato or a soiled glove the next. The year before there had been a cowboy: this year he was a hot potato, for he had become odiously familiar; a female palmist was also still in existence, but she was a soiled glove, since the pleasant *frisson* of having a bewildering future told in all the horror of detail before your friends, is an experience not to be repeated if subsequent events have shown the prophecy to have been altogether erratic.

But from the night of Lord Yorkshire's concert hope

began to wake in the season's middle-aged breast, that it, too, like most of its predecessors, would be known by an engrossing topic to mark it out from others before it was numbered with the colourless dead. For the picturesque—of a picturesqueness unequalled even by last season's cowboy—had at length arisen in the shape of Twins from Hampshire, Challoner Twins, Flints' nephew and niece. They sprang from a country parsonage, where Flints' brother, whom nobody had hitherto even heard of, lived like a sort of mediæval ascetic prophet, in a lugubrious atmosphere of fasting and prayer and scourging and sackcloth. He preached the most curdling sermons on Sunday, quite like Savonarola, on the comfortable doctrine of eternal damnation. About the twins, however, there was nothing in the least ascetic or mediæval: they were both quite young, hardly out of their teens, and were simply Diana and Apollo come to earth again. The girl (Helen too) had Titian hair in golden glorious profusion, a face like the morning and the inches of a goddess. And her charm, her bubbling spirits, her extraordinary enjoyment and vitality! She made everybody else look like a kitchen-maid, which was so delightful. But Martin, Phœbus Apollo, drunk with nectar! He played too: Karl Rusoff said he had never heard anything like it, and the dear old angel simply wept the other night at Frank Yorkshire's when Phœbus Apollo first dawned, but wept floods. And what could have been more romantic than the manner of their appearance? People were asked—we were all asked—to Lord Yorkshire's for "Music" in the bottom left-hand corner, expecting perhaps a couple of songs from Mattina and a nocturne of Rusoff's. Instead this divine boy walks up to the piano and plays the Pied Piper with us all. Yorkshire brought him up from the country, without a word to anybody and just shot him at London. He hit. Helen was with Lady Sunningdale—she always scores somehow—who gives out openly that she is madly in love with Martin, and makes him imitate her, which he does with such awful fidelity that it is impossible not to believe, if one shuts one's eyes, that it is not she who is talking. The only question is whether she will poison Sunningdale and insist on marrying Phœbus Apollo, or whether he will say "Retro Sathanas." It may be taken for granted that Yorkshire will marry the girl. Then the next night they were all four at the opera in Yorkshire's box, next the Royal box, and nobody looked at anything else. The girl was dressed in grey, very simple, but quite good. There was just a touch of blue somewhere, no jewels, but that radiant face and that glorious hair! Lady Sunningdale beside her looked like a lobster salad in the highest spirits. But really the boy is the handsomer, and when the opera was over people simply stood on the stairs to see them go out. But the twins were completely unconscious that it was they whom everyone was looking at, and came downstairs together, chatting, laughing and chaffing Lady Sunningdale because she had gone to sleep in the second act of "Siegfried." My dear, they are simply divine, and we must secure them at once for dinner or something, otherwise it will be too late.

The last sentence, whatever in this brief *résumé* of what London said was false or exaggerated, was certainly borne out by subsequent facts. For London, tired with its spinster ragelessness, rose at them as trout rise in the days of May fly, and besought their presence, finding them, as is not always the case with its rages, improve on acquaintance. They enjoyed themselves so enormously, and enjoyment is a most infectious disease, of which every hostess prays that her guests may sicken. They danced divinely, with the same childish pleasure, all night. Whatever the entertainment was, they were delighted, and their delight diffused itself through the crowds of which they were the centre. And it was always interesting to have at one's house the girl from nowhere who is going to make the match of the season, and the boy from nowhere who is going to send the world mad with music. The twins, in fact, blazed in the blue; they were the latest discovery, the point at which all telescopes were aimed. And they, presumably like the latest discovered star, were too busy to be either pleased or embarrassed that everybody was looking at them; they just sang and shone together with all the other lesser stars.

Ten days passed thus, Lady Sunningdale plying the bellows assiduously, and from time to time throwing on fresh faggots of interesting and picturesque information to feed the blaze. Nobody, not even the twins themselves, had been more astonished than she when they shot up into the zenith of success, for she had not anticipated anything of the kind, but that having happened, she was quick to assume the *rôle* of godmother. Nothing again, a week or two before, had been further from her thoughts than the idea that Frank Yorkshire should marry Helen; but that having been suggested to her, it was of course incumbent on her to say that she had brought them together with that express purpose, and, by dint of repetition, soon got to believe it.

The allied forces meantime had concerted their attack on that very well-garrisoned fortress known as Martin's father. Sheets of desultory letters were rained upon him by Lady Sunningdale which he answered with punctilious politeness, while Frank in far soberer strain told Lord

Flintshire the opinion of those like Karl Rusoff, who were thoroughly competent to judge, begging him use it in Martin's behalf. In consequence he wrote soon afterwards to his brother with some earnestness.

"You hardly ever come to town, I know, my dear Sidney," his letter ran, "but I really wish you would come now. It would make you prouder than you have ever been of both your children, if you saw them here. London—I am speaking quite seriously—has gone off its head about them. And, indeed, I'm sure I don't wonder. They are absolutely entrancing; their enjoyment of it all is the most infectious thing I ever saw, and we play ridiculous round games after dinner instead of grumbling at each other over Bridge. And their looks! Helen has taken the shine out of all the *débutantes,* and yet not one of them seems to hate her for it.

"This, however, is frivolous, but I want to tell you very seriously what an extraordinary impression Martin's musical abilities have made. He played the other night at Yorkshire's house, and I assure you all the musical lights of London simply hung on his hands. I know nothing about it myself; but when you find a great pianist and a great musician like Karl Rusoff listening, absorbed, to a young man of twenty-two, whom nobody has hitherto ever heard of, one cannot help attaching some weight to it. Others too, so Frank tells me, have been no less enthusiastic about him, but they are only names to you and me.

"Well, this is not entirely unasked advice, for I remember at Chartries a fortnight ago you consulted me about Martin and his future. And now it seems to me there is really no choice. He must be a musician; you cannot take the responsibility of trying to render unfruitful a gift like his. Nor would it be any good: he is bound to be one.

"Now, my dear Sidney, if there is any difficulty about expense—for I gather he must study exclusively for some time—pray do not give a thought to it. I will most gladly defray all expenses connected with him. Pray let me hear from you as soon as possible on the subject, and if you can run up to town for a day or two, you will see for yourself, and be a most welcome guest at the house of

"Your affectionate brother,

"FLINTSHIRE."

But in spite of these appeals and assaults, Mr. Challoner showed as yet no definite signs of yielding. To Lady Sunningdale his punctilious answers seemed mere frigid stupidity, and she had not the smallest or vaguest comprehension of the struggle that was going on in his mind. She could not understand that there was any choice to be made, still less that the choice could be a hard one, in determining whether Martin should once and for all close his dictionaries and open his piano, nor, had Mr. Challoner troubled to explain to her the deep mortification that Martin's ill-success in classical fields had given him, would she have been able to understand it. Karl Rusoff beckoned to him, and it passed her comprehension that his father should not, so to speak, throw him into the musician's arms. She could not, in fact, with all her acuteness, imagine in faintest outline any picture of the deep and real perplexity which Mr. Challoner was going through, a perplexity which for hours together tightened his mouth and ruled deep creases between the thick black lines of his eyebrows. The serious talks too which he had with his sister evening after evening between dinner and prayers and the temporary abolition of Patience would have seemed to her, if she had heard them, meaningless; they might as well have been conducted in a foreign tongue, of which she knew neither alphabet, grammar nor vocabulary.

One such occurred on the evening when his brother's letter arrived.

"I have heard from Rupert," he said, "who wants me to run up to town. That, I am afraid, is impossible. I have too much to do with Mr. Wilkins away for his holiday, and the Confirmation classes coming on. All the same, I should be glad if I could. His letter has troubled me rather."

"What does he say?" asked his sister, folding her very dry, thin hands in front of her.

"He says such extraordinary things; he says London has gone mad about them. They are amusing themselves enormously, it appears, at which, of course, I am rejoiced; but I can't help feeling a little anxious, a little nervous. They are so young, so thoughtless. I don't like the idea of people putting all sorts of foolish notions into their heads, making them think they are exceptional. I understand what people feel about them well enough. Dear children! I don't wonder at everybody liking them. But I gravely doubt whether it is the best of them that people find attractive, whether it is not their thoughtlessness, their unthinking high spirits, their looks, which attract others. That is so dangerous for them."

Clara Challoner put the pack of cards, which had been laid out ready for her Patience, back into their case. She did this without a sigh, because it was her duty to talk to her brother if he wanted to talk, and duty came before pleasure.

"That is exactly what I should be afraid of," she said. "The qualities that you and I, Sidney, were taught, and rightly, to consider weaknesses and blemishes, such as irresponsible high spirits and careless gaiety, seem to me now to be regarded as virtues. The younger generation shun earnestness and purpose in life as they would shun physical pain. Look at Lady Sunningdale, for instance, with whom Helen is staying——"

"Ah, give her her due," said Mr. Challoner; "she is a very clever woman."

"But to what does she devote her cleverness? To the mere pursuit of frivolity. I wondered, as I told you before, whether you were wise to let Helen go under her wing. She will be among people whose only aim in life is amusement. That is the one thing they take any trouble to secure."

Mr. Challoner shook his head.

"I hope, I pray I have not done wrong to let them go," he said. "I did it with a definite purpose, in order to let them see that sort of life. Helen is not naturally frivolous. Look at her work here with her classes, how admirable she is, how they adore her, how her heart is in it. And to bring a girl up in ignorance of what the world is like does no good. Sometimes I wonder whether I have not sheltered her too much, kept her too much in this sweet place with all her duties and pleasures round her. But it is not of her that I am most thinking. She will come back unspoiled, with just the memory of a great deal of laughter and innocent amusement. No, it is of Martin. Rupert speaks chiefly of him."

He took the letter he had just received from his pocket, and read it to her.

"It is a great puzzle, a great difficulty to know what to do," he said. "Even at Cambridge, where he is surrounded by all those grave industrious influences, Martin does not seem to me to gain in depth, or in set purpose of life. And if I consent to this, he is plunged into surroundings that so much more conduce to shallowness, to indulgence of the senses. Thank God, I believe my son is pure. But he is so impressionable, so easily stirred by enjoyment into thoughtlessness, that I am very much afraid."

He got up and moved over to the window, where he stood looking out. In front the ground sloped sharply away down to the churchyard, where in the last fading light of evening the grey tower stood, like a shepherd watching over its flock, among the gravestones, below which rested the bodies of those entrusted in sure and certain hope to its hallowed care. Like all strong, hard-working men, Mr. Challoner was far too fully occupied in the duties of this strenuous life to give much thought to death, except as to some dim quiet friend whose hand some day he would take without fear or regret. But how terrible death could be, and how terrible it would be to him, if through carelessness or biassed judgment he had chosen wrongly for one so dear to him, so peculiarly entrusted to his care! How terrible again would be that quiet friend, if, through want of wideness in sympathy, he had tried to nip, to starve, to stifle a gift with which God had endowed his son!

Then suddenly with a wave of bitterness all that he had planned in long sweet day-dreams years ago for Martin filled his mind as the harsh salt-water fills a creek. He had seen him a scholar, minute, painstaking, absorbed, perfectly himself in accuracy and subtlety of mind by the study of the great classical authors. He would be a fellow of his college; and his father, so he pictured to himself, would live over again his own college days, which perhaps were the happiest in his life, when he saw Martin seated at the high table among the masters of learning, or in professional gown crossing the dear, familiar grass of the quadrangle to the grave grey chapel on summer afternoons when the sun made jewels of the western panes, or in winter when the soft mellow glow of candles shone dimly by the dark oak stalls, and scarcely reached to the vaulted fans of the roof.

Then the picture took larger lines. With the wealth and position that would one day be his, there was no limit to the influence that Martin might have in an England which even now seemed to Mr. Challoner to be dozing in a stupor of contented, unreligious, unintellectual enjoyment. There was need of a scholar, a man with a great position, a man of strong Christian faith, to arise, who, with a life unselfish in its aims, liberal, charitable, encouraging all sorts of godly learning and scholarship, should give to the world a strenuous intellectual ideal again. How often in his prayers had that vision risen before him, that future which he desired so eagerly for his son, and which, so he believed, was humanly possible for him? Chartries should be again what it had been four generations ago, the centre of the scholarly intellectual men of England. The accounts of those days in the history of Chartries read to him like a wonderful true fairy story. Three or four times in the year the house was filled by his great-grandfather with men of learning, and after breakfast and morning service in the chapel they would meet and discuss till dinner-time some exquisite point of scholarship, or hear from some expert of the latest discoveries in the Roman forum. At these discussions his great-grandmother, a woman of great culture and knowledge, had always been present: she had once even read a paper on the Elgin marbles, then but lately come to England, in which, with a marvellous subtlety and accuracy of observation, she had upheld the view, in the face of strong attack, that they were Greek originals, not Roman copies. This and all other papers read there were preserved among the printed "Horae aureæ Chartrienses," which was the record of these gatherings.

For Martin then he had dreamed a life like that, the life of a cultured, scholarly, Christian gentleman, not monkish, but with a brood of growing children round him, busy at his books, busy in all matters of education, instant in prayer, and a churchman staunch to uphold the rights and the glory and the privileges of the mother of his faith. Instead, he was asked to give permission that Martin, after years of expensive education which had ended in utter failure, should devote himself to music, or as Mr. Challoner put it to himself, to "playing the piano," a profession which to his mind was akin to a sort of mountebank's. Nor was that all; if it was only in intellectual attainment that Martin had shown himself desultory and idle, his father would, it is true, have deeply regretted it, but it would have been as nothing compared to the anxiety he felt with regard to that slackness and indolence of character which he thought he saw in him. Left to himself, he would lounge the day away, not only without acquiring knowledge of any kind, but without a thought as to the strengthening and building up of his own character. He would scribble amusing sketches by the score, play on the piano by the hour, or, as like as not, lie on the grass and smoke, in purposeless waste of these infinitely precious hours of youth. Had he ever shown interest in matters naval, military, or political his father would gladly have seen him a soldier or a member of Parliament. But he was purposeless, desultory, without aims or interests, and so utterly unlike himself in every point of character that he could scarcely believe he was his son. And this estimate was no new one; ever since Martin was a little boy, through his school life and through his four years at the University, he had noticed the same drifting weakness, the same tendency to take any amount of trouble to save trouble. Nothing had made any impression on him, not his Confirmation, nor his growing responsibilities as he rose in the school, not the duties attaching to the sixth form when he was dragged up into it, nor the widened life at Cambridge. It was all one to him; he had the pleasant smile when things went well, the yawn when effort was demanded of him, the eternal drifting towards the piano.

All this passed through his mind with the rapidity of long and bitterly familiar thought.

"They all urge me to do it, Clara," he said; "yet they don't know him as I do, and they are in no position of responsibility with regard to him. I can't see my way at all. It is no use his continuing to waste his time at Cambridge, and yet London for my poor rudderless Martin! What influences may he not come under? Who is Rusoff of whom Rupert speaks? But I must settle; it is no use putting off a decision that has to be made."

He turned away with a sigh from the window.

"In any case he had better come home for a day or two before he goes up to Cambridge," he said, "so that

I can talk it all over with him. In fact they had both better come home. They have been in town a fortnight, a fortnight of pure amusement. Besides, the parish library wants looking after."

"I can manage that if you would like Helen to stop a little longer," said his sister.

"No, dear, your hands are full enough already. Besides, Rupert's letter has made me altogether a little uneasy. It is time they both came home."

CHAPTER VI.

HELEN was seated at a big plain deal table in the village room, with a large array of volumes in hospital spread in front of her. Some wanted covers—the cover peculiar to the books in the library of the room was brown holland of a strangely discouraging hue, stitched over the back and sides, and turned down inside—others wanted stamp paper over torn edges, and most wanted labels, bearing the title, gummed on to their backs. True, the very magnitude of the repairs needed was evidence that the library was at any rate appreciated by the parishioners; but the thought that her nimble hands were employed on a useful work did not at the present moment succeed in consoling her for the extremely distasteful nature of the occupation. Dispiriting too were her surroundings: on the walls hung the hateful maps of Hampshire and the Holy Land, scientific diagrams of the construction of flowers, and several charts of geological strata showing old sandstone, new sandstone, blue lias in which diamonds occur, and yellow bands of auriferous reef. A large black cast-iron stove stood in one corner, a bagatelle board with torn cloth and tipless cues—these too would have to be mended after the library—occupied another table, and standing against the wall were low deal bookcases. The floor was covered with an affair of oil-cloth pattern and of corky texture, so indestructible as to be practically eternal; and a harmonium, happily not at all eternal, but in advanced senile decay of cypher and dumb notes and strange noises like a death rattle, stood near the door. In spite of the wide open windows, the characteristic smell of the room hung heavy and stale on the air in this oppressive heat of an August day.

Helen had been back from London some three weeks; but in spite of her endeavours to settle down again into village life, she had not been very successful in doing so. Duties which before had seemed tolerable enough had become frightfully tedious, while those which before had seemed tedious had become intolerable. Only the evening before her father had spoken to her about her general behaviour *apropos* of what he called the "falling-off" of the village choir. This meant that on the previous Sunday the organist had played one tune, and the choir had sung another, which had displeased his unmusical ear; though Martin, who had been home from Cambridge for the Sunday, had listened with rapt attention, and said to Helen that he thought it extremely Wagnerian. This opinion, it may be remarked, he had not expressed to his father.

"I am afraid your pleasure trip to London has unsettled you, Helen," her father had said, "and you should really take yourself in hand, and make up your mind to recapture your habits of industry again. One is often disposed to be impatient with what one calls 'little duties,' but, dear girl, there is no such thing as a little duty. There is no such scale possible; duty is duty, and it is all great; and your eager and willing performance of all those things which may seem to you small is just as much a part of real life as to the emperor the discharge of the cares of his empire. For instance, the hymn at the morning service on Sunday——"

"But it isn't my fault if Mr. Milton plays the wrong hymn," said Helen.

"But it ought to be impossible that such accidents should occur," said her father. "You should think, dear Helen, in Whose honour we stand up to sing in church; and that knowledge constantly with you, you will find must elevate the smallest duty, and raise the most insignificant piece of work into an act of praise and worship."

"I will try, father," said she.

"I know you will. Your holiday, all the mirth and innocent pleasure you have had in London, ought to help you to it. Those times of refreshment are given us not to make us discontented with our work, but to enable us to bring to it a rested and more active industry."

But this morning it seemed to have brought to Helen nothing of the kind, but only a rested and more active doubt as to whether any of the things that filled her day could possibly, in the doing, be good for her, or, when done, for others. The *Sunday Magazine*, for instance, of which at this moment she was pasting the torn pages, seemed to her to be singularly ill adapted to do anything for anybody. There was an essay on the habits of mice; another on the temptations of engine-drivers; answers to correspondents dealt with lotions for the hair and the best treatment for burns; while in the forefront of each number was an instalment of a serial story connected with incredible ranches and unlikely mining in California. But, in spite of her conscientious doubts, her fingers moved apace, and the stack of healed and mended volumes at her right hand grew quickly tall.

She worked on till about twelve without pause, and then pushed back her chair, and began carrying the mended volumes to their shelves. If only she could have entertained any hopes as to the utility of what she was doing she would have accepted her occupation with cheerfulness, for her nature was one of that practical kind which finds almost any pursuit, so long as it has definite and profitable aim, congenial. This afternoon again she would have to take choir-practice in the Room, and even with the eager desire to find "good in everything," she could not see who profited by the cacophonous result. And, to add to her labours, the ill-inspired ambitions of Mr. Milton had caused him to learn with infinite pains and groanings of the organ in evening hours nothing less than an anthem for the harvest festival, and it remained for her to teach the choir. Hours would go to the repetition of it before that unmelodious festival, and even if it had been possible that relentless practice could make the choir tolerably secure of their notes (which it could not), yet the result, even if it were faultlessly performed, would be deplorable, since it was an anthem of that peculiarly depressing kind produced by minor organists, which contained a fugual passage which was not a fugue, and, musically speaking, was of the most suburban and jerry-built construction.

Helen pushed back her hair, and slightly amused at the greyness of her own thoughts, smiled to herself as she went backwards and forwards between table and book-case. If only she had someone—another sister—to share in these farthing woes of a rector's daughter, she could have laughed at them; she and a friend at any rate could have read each other striking extracts from the mended leaves of the *Sunday Magazine.* Then suddenly she heard a step on the gravel outside—a step, not her father's, but strangely familiar, and the door opened.

"Why, Lord Yorkshire!" she said. "How delightful! Do come in."

Frank had the enviable faculty of keeping comparatively cool on very hot days, just as on occasions heating and stirring to the spirit, his nature seldom boiled. But to-day he was much hotter spiritually than physically, and Helen's genuine pleasure to see him, which shone in her eyes and her smile and vibrated in her voice, did not reduce this genial heat.

"I have not done wrong," he asked, "to come and interrupt you? They told me at the Vicarage that you were here."

"No, indeed you have not," said she, shaking hands. "Really, I was longing for an interruption. Look!" and she pointed to the titles of her mended stack of books.

He glanced at them with a smile.

"Really, without undue conceit, I don't wonder," he said. "And so this is the Room you told me about in London."

His eyes wandered round, looking at the maps and the coloured chart of geological formation, at the harmonium, the bagatelle-board. Then suddenly all the girl's loyalty to her father rose in her.

"Ah, don't laugh," she said. "I can't bear that you should laugh."

He looked at her quite gravely.

"Heaven forbid!" he said. "Here, as in that map of geological strata, there is an auriferous reef. There is to be found a little belt of gold in everything which we may have to do, as long as it is not—not nasty. The trouble sometimes is to find it. Haven't you struck it this morning?"

Helen sat down with a little sigh.

"No. Help me to dig a little," she said. "Look at the soil! *Sunday Magazine.* A serial. Then 'Round the tea-table," with a receipt for muffins. 'Muffins' is torn, I must mend it. Missionary work among the aborigines of Somaliland. Oh, dear! What has it all got to do with me—this me?" she cried.

"Perhaps you have not yet mended enough to find out," he suggested.

"That is possible. All the same I have mended a good deal. Now I am going to talk 'Lady Sunningdale' for two minutes: at least, there are fifty distinct and separate things I want to say in one breath. First of all, please smoke; the room smells of Sunday school. Yes, and give me one. If my father appears suddenly you must say it was you. Next, I suppose you have come from Fareham. How is Lady Sunningdale? And you'll stop for lunch, of course. Next, Martin; he is going to leave Cambridge at the end of the Long; he is going to settle in London in the autumn, and study under Monsieur Rusoff. Oh, why wasn't I born a boy? I suppose you can't tell me? So once again about Martin, thanks. What a good time we had in London! I have never enjoyed a fortnight more. Is everyone as kind as that always?"

"I think they always will be to you," said he. "You two took London by storm. We all went into mourning and retirement into the country when you left."

Helen laughed.

"You don't look as if any grief particularly weighed on you," she said.

"Clearly not now," said he.

This was a little clumsily obvious, and it made her for the moment slightly embarrassed. She dabbed a label somewhat crooked on to the back of a work about missionary enterprise.

"Can you write a legible hand, Lord Yorkshire?" she said. "If so, and if you will be kind enough, please write *Sunday Magazine* very clearly on twelve labels, with ordinal numbers, one to twelve, below the title. And when I've pasted them on, I shall have finished, and we will go out. Martin isn't here, I am afraid. He is up at Cambridge till the end of the month."

Frank obediently took a pen. He had suffered a slight repulse.

"A notable charm of life," he remarked, "is its extreme unexpectedness. If I had been told by a chiromantist that I should shortly be writing the words *Sunday Magazine*—is that legible enough?—twelve times over with numerals beneath, I should have distrusted everything else he said. Yet, here we go."

Helen laughed: she was not quite certain whether she was pleased or not at the success with which she had turned the conversation on to topics so alien from herself as the *Sunday Magazine.*

"Quite so," she said. "And if I had been told that I should be telling you to do so, I should have considered it too wildly improbable to be even funny. Yet, as you say, here we go. Oh!"

Her ear had caught the sound of a step outside, and with a quick sweep of her arm she threw her cigarette out of the window.

"It's you, remember," she said with whispered emphasis.

Frank's cigarette, however, was still unlit, but he obligingly remedied this, and hurriedly blew out a cloud of smoke and silent laughter. Next moment the Vicar entered. He paused for a second on the threshold, his nostrils surprised by this unusual aroma in the Room, but Frank instantly rose.

"How are you, Mr. Challoner?" he said. "I called at the Vicarage, but everyone was out. But hopes were held out to me that I might find some of you here, so I came. And behold me," he added rather felicitously, "a lay helper," and he pointed to his half-written labels.

The Vicar's somewhat grim face relaxed. There was a neatness about Frank's speech which his classical tastes approved.

"It is too kind of you, Lord Yorkshire," he said. "Helen has impressed you into the service, I suppose. But—I am sure you will excuse me—would you mind finishing your cigarette outside? Our rules about smoking in the Room are stringent. You will excuse me."

His eye glanced rather sternly as he spoke at Helen. This was one of the laxities he deplored in his children. She knew quite well that smoking was not allowed in the Room. The most infinitesimal moral courage on her part could have stopped it. And he himself knew how she would excuse herself, saying that she did not think it mattered in the morning when there was no one there. It was a rule of the place, however. He had made it; she knew it.

Frank instantly threw his cigarette out of the window.

"I am so sorry," he said, "and I am afraid I never asked leave."

"You have no idea what difficulties we have with even quite the small boys of the village," continued Mr.

Challoner. "Children of eight and nine think it manly to pull at an inch of bad tobacco. So I am sure you will not even mentally accuse me of faddiness. I gave up smoking myself entirely for that reason. You are too kind to help my daughter. You will lunch with us, of course."

"Thanks very much. I came over in the motor from Fareham, and Miss Helen has already been so good as to suggest ——"

"Of course. Martin is away from home, I am sorry to say. Helen has no doubt told you what has been decided."

He glanced again at her as her quick nimble fingers plied the work which an hour ago had seemed so distasteful. Certainly now there was in her no trace of that listlessness and want of application and vitality that a few days before had occasioned his loving rebukes. She was all vivid and alert: the fresh bright colour shone like a sunlit banner in her cheeks, and as he looked, he realised for the first time that this was no longer "my little girl," but a woman in her own right. Then, like an echo to this, came the thought that he was not the proprietor of his children. Adviser, corrector, pruner, cultivator he might be; but he could not make nor stop growth, if "my little girl" decided otherwise.

This was something of a shock, though only momentary, and there was no perceptible pause before he spoke again.

"So you will bring Lord Yorkshire home to lunch, Helen," he said. "I must go on to the village. I only looked in on my way. Half-past one, Lord Yorkshire. And afterwards you must try a cigar that I can give you. A year ago they wanted keeping, and now they have got it."

For a little while after he had left neither spoke. A label had been put on crookedly and required re-adjustment: something else also had gone crookedly, and Helen had to re-adjust that too.

"I'm afraid I must tell him I had been smoking," she said. "Oh dear, what a bore."

"Is not that too transcendental honesty?" he said.

Her eyes flashed their wide light into his.

"Ah no, there is neither less nor greater in honesty," she said. "It is a great bore to be honest. I wish I wasn't. No, I don't wish that. It is one of the uncomfortable things which one can't get on without."

Suddenly he knew that a moment, which for weeks had been approaching slowly, rushed into the immediate future. He sat upright in his chair and quite unconsciously moved it nearer hers. His upper teeth closed on his lower lip, dragging it upwards till it was white. Some mad current of blood sang in his ears, some sudden mistiness obscured his eyes, and she was but a dim, wavering form close to him.

"Honesty! honesty!" he said, "Helen!"

A long-drawn breath rose in her bosom, filling it, filling everything. A *Sunday Magazine* dropped from her hand, and she stood up. He too stood, and they faced each other for a long moment, and the new certainty became the only certainty there was.

"Oh, are you sure, are you sure?" she cried.

And there was no more need of words just then.

"Since you took the hare out of the trap," she said. "I think I loved you for that."

"Since you caught my finger in the trap."

"And it bled," said she.

"But you bound it up for me."

She raised her face and held him by the shoulders, arms outstretched.

"And I remember saying to Martin that this was the sort of room in which nothing nice could happen! Oh, Frank, how has it happened? How has it happened?" she said.

"I don't know how, my darling, but I know why."

"Why then?"

"Because it had to happen as far as I was concerned. Because it was you, in fact. How could it have been otherwise?"

Her eyes dropped a moment, and then looked full at him again.

"Is it real?" she asked. "And if it hadn't happened, what would have become of us? Supposing you had not said 'Helen'?"

"What else could I have said?"

"You might have said nothing."

"Nothing? You and I here together, and nothing? I had been saying nothing too long," he cried.

"No, not too long. It has all been perfect. And—and the *Sunday Magazine,* and, and twelve labels each with their numbers. Oh, I surrender," she said.

"When you have utterly conquered?"

"Yes, both. And both of us."

"There is only one."

It was no descent to return to the unfinished work; the business of label-pasting rather was illuminated and made glorious, the putting of the books back in the shelves was a procession of love. Then came the return to the Vicarage under the benediction of the sun, and the intrusion of the presence of others; but as some telegraph from lover to lover throbs across hundreds of miles of arid and desert country that does not know what secret and blissful tenderness has passed over it, so from each to the other passed unnoticed glances that sent the electric current to and fro, and the words of common life were to them a cypher charged with intimate meaning.

It had been settled between the two that her father should be told at once; and accordingly after lunch, when he went into his study to get Frank the promised cigar, with a view to coffee on the shady croquet lawn, the latter followed him, while the two ladies went out, and told him.

"It is the happiest day of my life, Mr. Challoner," he said very simply. "Your daughter has accepted my devotion and love."

Mr. Challoner turned to him quickly.

"Helen?" he said—"You? Lord Yorkshire, this is most unexpected. But I am charmed, delighted at your news. And I risk the imputation of a father's partiality, when I say that I congratulate you most heartily."

He shook hands warmly with the young man, and an emotion very deep and heartfelt vibrated in his voice.

"May the blessing of God be on you both!" he said.

For a single moment Frank felt as if the thermometer had dropped suddenly; but the sensation was so instantaneous that before he could analyse it, it had passed, and Mr. Challoner still held his hand in his strong firm grasp.

"And I think, I believe, she is a very fortunate girl," he added. "When, when did you speak to her?"

"This morning only. We settled to tell you at once."

"Thank you. That was right of you. How the years pass! Why, it seems only yesterday—well, well—let us join them outside. Ah, a cigar for you. I declare I had forgotten."

They crossed the lawn together, and as they approached the group of chairs underneath the box hedge, Mr. Challoner quickened his step a little and advanced to Helen with hands outstretched.

"Helen, my dearest girl!" he said.

The glorified hours of the golden afternoon passed too quickly. Parish work soon claimed the Vicar, who, as he passed through the village, gave notice in the school that the choir-practice was postponed till the next day; Aunt Clara betook herself to district-visiting, and the two were left alone again while the shadows began to grow tall on the grass. Sweet words and sweeter silences sang duets together, and from talk and silence they learned each other. For their falling in love had been an instinctive, inevitable thing; and now that the gracious deed was accomplished, they explored each other's nature in the excellent brightness of the love-light.

"Lazy, frightfully lazy!" said he. "Will you take that in hand for me? With the unaccountable delusion, by the way, that I am extremely hard-worked. I lie in bed in the morning, and groan at the thought of all that I shall have to do before I go to bed again. After a very long time I get up—and don't do it. Helen, how could you have been in the world all these years, and I not know it?"

"Oh, what does it matter now? For here we are, and for all the rest of the years we shall both know it. Yes, you shall get up at seven every morning. I will wake you myself."

"That will be nice. And I needn't get up at once? And what am I to do when I do get up?"

"Why all the things you lie groaning about," she said.

"But there aren't any really. At least nothing to groan about."

"Now you're talking nonsense. I don't mind, though. You talked a good deal of nonsense on that Sunday, the hare-Sunday, you and Lady Sunningdale. How is she?"

"I forget. I forget everything but—this!"

She bent towards him.

"Am I really all that to you?" she asked.

"Yes, all. More than all."

After awhile she spoke again.

"And you have no back-thought? There is no dark place at all. No shadow of any kind?"

He looked up quickly.

"Yes, a possible shadow," he said.

"Religion?"

"Yes; it had occurred to you, too. What do you expect?"

Helen sat with her chin resting on her hand a moment without replying.

"I don't know," she said at length. "Don't let us think about it just now, Frank. Let this afternoon be perfect. But I can tell you this, that though it may possibly be very painful, it will make no difference to me. I shall be very sorry—very, very sorry, but—— That 'but' is you, if you understand."

"Thank you, my darling," said he.

Mr. Challoner carried a very thankful heart with him as he went on his various errands that afternoon. To see Helen happily married was a constant desire and prayer of his; and though he would with willingness and thankfulness have given her to the keeping of any good man who could support her and a family, he did not attempt to disguise from himself the satisfaction he felt at her having made what is vulgarly called "a great match." She had the gifts which should enable her to fill a great position, and to play a great part worthily was a bigger and a finer thing—though, as he had said, "duty was duty," and there is neither less nor greater—than to work on a smaller scale. More than that, he had, with all his personal unworldliness, a good deal of pride of race, which Frank, with his undeniable birth and breeding, gratified. For the man himself also he felt a very decided liking and respect. He was an admirable landlord, in spite of his avowed laziness; he was generally considered to get through the day's work with credit. In the House of Lords, also, he had already achieved a certain reputation for eminent commonsense; and though to advocates of extremes his speeches might appear commonplace, that was rather the fault of those who held an extreme view. In other words, he lent his wealth and position to the support of moderation, much as Lord Flintshire had done.

Another matter, dearer to Mr. Challoner's heart than the obscurities of fiscal affairs, was that Frank was, if not a pillar, at any rate a very sound piece of the fabric in the twin-towered building called "Church and State." His patronage was always given to clergy of moderate views, who did not indulge in what Mr. Challoner called "idolatrous and Romish practices;" while, on the other hand, he always voted dead against any attempt to subtract from the power or position of the English Church as by law established.

"A staunch Churchman," said Mr. Challoner to himself, as he walked with his long rapid stride through the pathway hedged about with the yellowing corn.

For the time his disappointments about Martin were

forgotten. There, it is true, his dreams about his boy's future had been dispelled by a rude and bitter awakening; but here, at any rate, was something which he had never dreamed being realised; and without overestimating the force and value of education, and the influences which spring from environment and mode of life, he believed that Helen would assuredly live her mature and wider life on the lines in which she had been brought up. So in this marriage he saw a strong weapon forged of steel and wielded by a loyal hand in defence of his mistress the Church. He knew well the immense power which in England a territorial magnate is possessed of, how by the mere fact of his wealth and position he can control the course of wide issues. Hitherto Frank had done just that; he had always ranged himself on the side of education and religion, or rather he had ranged the inert weight of all he represented there, while he himself had keenly pursued the artistic things of life. But now Helen, with all the influence of her home and upbringing strong within her, would come to add life to this solid weight, making it an active and potent instead of a passive instrument of good. He almost envied the girl; such opportunity was given to few only, and on her would the responsibility and the glory rest.

His district-visiting that afternoon had taken him into the farthest limits of his parish, and a three-mile walk into the glories of the sunset lay before him when he turned homewards. A flush of colour, vivid and delicate as the cheek of youth, incarnadined the west, over which a few light fleeces of crimson cloud hung like flames, and further up from the horizon a belt of aqueous green melted into the transparent blue of the sky overhead. The sun had already sunk behind the tawny line of swelling down, and the water-meadows by the Itchen, where his path lay, were full of dusky and deepening shadows. Right down the centre ran the lucent stream, reflecting on its surface the blue and the green and the flush of the sunset sky. Rooks cawed their way homeward to where the elms of Chartries showed black against the luminous west, and to the left of the long gabled façade of house-roof rose the grey gothic tower of his church, the lodestar of his life, the mistress of his heart. That was the most real thing in all the world to him; all that was beautiful at this magic hour in earth and sky was but a path that conducted his soul thither; all that he loved on earth was only the shadow and faint similitude of the great love of his which centred there. Nothing had any true existence except in its relation to that, everything else was but an avenue to an antechamber in the house of many mansions. And as his eye first caught sight of the grey cross-surmounted tower he stopped a moment, uncovered his head, and, with closed eyes, stood still in a Presence more poignantly there with him than any. Through his impatience with ways and methods not his own, through his intolerance of that of which he had no ability of comprehension, through his instinctive dismissal of all that seemed to him unessential in life, whether it was the benediction of the evening hour, the piano playing of Martin, the sweet eyes of Helen, through all, at moments like these when his human emotions where most aroused, his view pierced triumphant, and saw only the Cross of Christ pointing heavenwards. Towards that and that alone the essential nature of the man was directed, even as the compass-needle, though deflected and distracted by other neighbouring agencies, is essentially undeviating and loyal in its allegiance to the north. His disapprovals, his censorious judgments, his want of sympathy for what he did not understand, were only the husk of the man, and it was the very strength of his central devotion that made him intolerant of any who seemed to lapse in things great or small from his own measure of fervour. Extreme cases, indeed, the case of the Jew, the Turk, the infidel, he left with faith to the mercy of God, though his human comprehension did not see how they could be capable of receiving it. He did not know; he left them before the Throne of Infinite Compassion, and turned his thoughts elsewhere, to his own work of ministering to the sick and needy, to the cultivation of the intellect, the usury of that sterling talent given to man, and all that should make a man more capable of worship, a fitter instrument in the hand of the Great Artificer.

The rose colour in the west faded to the nameless and indescribable hue of the hour after sunset, a single spangle of a star flashed in the vault of velvet sky, and dusk, like the slow closing of tired eyes, fell layer after layer over field and copse and river. Lights began to twinkle in the cottages of the village, day with its joys and its work and its rewards was over, and rest was ordained for the world and its myriads. Instinctively the mood of the tranquil hour gained on him, his foot abated a little from the vigour of its stride, the active fervour of his brain cooled a little, and a very human tenderness rose and suffused his thoughts. Here in the churchyard which he was now crossing stood the plain marble slab, with its lettering now twenty-four years old, below which lay the remains of her who had been the one passion, short and sweet and bitter, of his life. How often in those years had he wondered, with aching longing for light, what was the design of that interlude, what was the correct reading, so to speak, of the passion that had for a year so absorbed and mastered and overwhelmed him! His wife and he had no spiritual affinity; his love for her had not raised and inspired him, and he, strong and loving as he had been, had not helped her with any success towards the strenuous and active service which he knew to be the bounden duty of every living soul. Had his passion then been merely a casual carnal longing, a frailty of the flesh? Often and often he had been afraid to answer that question honestly; but to-night, as he paused for a moment by the grave, that doubt assailed him no longer, and instead, a strange yearning and regret for a missed opportunity took its place. Had he dealt wisely and gently with that sunlit child? Had he failed to realise what a child she was, and been harsh and deficient in tenderness to a little one?

His head drooped for a moment as he stood there, and then with all the honesty of a nature as upright as a fir-tree he answered it. He could not justly condemn himself; he had done his best according to the light that was given him. He had acted in a way he would have advised another to act—he would act so again now. It had not been easy: often he had longed to kiss her face into smiles again, and had been stern instead.

Then briskly again he left the grave, and in the gloaming stepped across the lawn into the long window of his study. The lamp was already there, trimmed and lit, his work was spread on the table in orderly array. There were still ten minutes remaining to him before he need dress for dinner, and from habit long-engrained he sat down at once to use them. He found his place, composed his mind to the topic on hand, and dipped his pen in the ink. But contrary to habit his attention wandered, and strayed back to the churchyard, and until the dressing-bell sounded, he sat there looking out of the window with unseeing eyes, questioning, questioning.

CHAPTER VII.

THREE glasses of claret during dinner and one of port with his dessert was Mr. Challoner's usual allowance of alcoholic fluid, and as a rule neither his sister nor Helen took any. But to-night in honour of the occasion a half-bottle of champagne to drink a toast in which two names were coupled made its unusual appearance, and the Vicar proposed the health in a voice which shook a little with feeling.

"God bless you both, my dearest girl!" he said, and drained his glass.

Afterwards, as if to endorse the felicity of the occasion, the malignancy of the cards was abated, and Aunt Clara's Patience "came out" twice before prayers without a semblance of cheating on her part. Why she cared to play at all if she cheated had long been to Helen an unanswerable riddle, and was so still. But in her dry and passionless way to get out without cheating was a satisfaction to Aunt Clara. She was pleased also with the engagement of her niece, but her comparative reticence on that, as on the subject of Patience (she had said only, "Fancy, Sidney, Miss Milligan came out twice!"), was due not, as in her brother's case, to excess of feeling, but to the inability to feel anything at all acutely. The performance of her duties in the house and in the parish had been for years a sufficient emotional diet; from other influences, like a freshly-vaccinated person in respect of smallpox, she was immune. She always said "good-night" the moment prayers were over, and did so on this occasion. But she kissed Helen twice. That corresponded to her observation to her brother about the "Patience."

To-night, however, contrary to custom, the Vicar lingered in the drawing-room instead of going back to his study, and when her aunt was gone, Helen took this opportunity of getting her little confession made. He had beckoned her to the arm of the long deep chair in which he was sitting, when she would naturally have followed her aunt upstairs, and took her hand in his, stroking it softly. Such a spontaneous caress was rare with him, and, in spite of the enormity of her confession, she needed no large call on her courage to make it.

"There is one thing I want to tell you, father," she said. "I hope you will not be very angry with me."

Mr. Challoner pressed her hand gently. Now, as always, the confidence of his children was a thing immensely sweet to him; to get it unasked, pathetically so.

"What is it, dear?" he said. "I don't think you need be afraid of that."

"Do you remember this morning asking Lord Yorkshire, Frank, not to smoke in the Room?" she asked.

"Yes, perfectly. And since I feel sure I know what you want to tell me, it did occur to me that you might, with a little courage, have asked him not to. You knew my feeling about it. But you have told me of your own accord, dear. So that is finished, quite finished."

The temptation to say no more was extraordinarily strong, and to end this beautiful day quite happily with everyone—Aunt Clara had kissed her twice, which she usually only did on Christmas morning—was the childish impulse dominant in her. To-morrow she would deal with other things; and one perfect pearl of a day would be hers, an imperishable treasure. But the necessity of honesty, consecrated, as it were, by what had passed between her and Frank on the subject, conquered. For the last year she had occasionally smoked, and had never in the least desired to tell her father that she did. Yet now, somehow, perhaps because it was connected with him, she must. So she spoke:

"No, it is not quite finished," she said. "I had been smoking, too."

For a moment he almost failed to grasp this simple statement, then a schoolmaster-voice rapped out a question:

"You smoke?" he asked.

"Not often, not much," she said, with the old childish awe of him suddenly returning.

"And who—did Martin teach you?" he asked, with an ironic emphasis on "teach," at that fine word being put to such base uses.

"No, at the beginning I asked him for a cigarette," she said.

"And he gave it you?"

There was no reply necessary. He had dropped her hand as if it had been a cigarette-end, but now he took it again.

"My dearest girl," he said, "I do not want you for a moment to think that I make much out of a little; do not think that I regard it as morally wrong in any way. But think, Helen; a girl like you smoking—is it seemly? Is it not a horrid, a nasty habit? And in the Room, too! There, there, don't tremble, my dear. I am not angry."

There was a moment's pause.

"Let us dismiss it altogether, Helen," he said. "You

told me, anyhow, and I know it was hard for you to do that. But"—and he was father, responsible father, when he should have been friend—"but you knew my feeling about it. It was disobedient."

All the time his heart was warmed by the thought that she had told him; yet his sense of duty, his responsibility towards his children, which was one of the most constant motives of his acts, made him say more. He did not want to preach, but he was incapable of not doing so.

"Yes, disobedient," he said, "to what you knew I felt. And that Martin should give you a cigarette is as bad."

"Ah, do not bring him into it," she said. "I am stronger than Martin—he had to give it me. Martin would always do what I asked him. Please do not write to him or speak to him about it."

Then at the thought of Martin, and of the constant, continual misunderstandings between him and her father, her own great happiness urged her to try to help him.

"I am much worse than Martin is, dear father," she said, "much more disobedient, much—'The Mill on the Floss,' for instance. I had been reading it."

"And he had lent it you?" asked Mr. Challoner quietly.

"No. I found it in his bedroom and took it. Oh, father——"

The issues for each had deepened. The meaning of that exclamation was understood by him: it pleaded with him for Martin.

"I have always tried to be a good father to you both," he said.

Then all that Helen had suppressed and striven not to have thought for years rose to the surface on this her first day of liberty. She had not let herself know how heavy the yoke had been till now, when her manumission was signed. But Martin still was in subjection. She stood up.

"I know that," she said. "If I had not always known that I should not have cared. It is just that which makes it so sad. But we have both been afraid of you. We have concealed things from you because we were afraid of your displeasure. You know Martin is awfully timid! he shrinks from what hurts. And we do not tell you everything even now."

The thrill of pleasure that her unasked confidence had given him had pretty well died out. He felt also that there was something more coming.

"You or Martin?" he asked.

The tide was irresistible, sweeping her away. A thing which must be horribly painful to him had to be told her father to-day, to-morrow or some time, and she suddenly knew that she must tell him now. Besides, here was a burden she could voluntarily bear for her lover, a pain, a difficult thing she could take on herself. And, woman all through, as she would have saved him anything from a toothache to a heartache, especially if the saving it from him meant the transference of it to her, she felt, in spite of the pain, an inward thrill and warmth at the thought that it would be spared to Frank. A few minutes before, when Aunt Clara left the room, she would have gone too, if she had known that the little confession would lead on to this, but now the burning of her love, as when a furnace door is thrown open, glowed with a whiteness that consumed all else.

"I, anyhow," she said. "I have something which you must be told. And I choose to tell you instead of Frank."

Her father got up also, facing her. He was very grave, very still.

"Does it concern him?" he asked.

"Yes."

"Is it disgraceful?"

"No."

He made one futile attempt to stop in the middle of the rapids into which he or she, he did not know which, had steered.

"Then tell me nothing, Helen," he said. "You say it is not disgraceful. That is quite sufficient for me when it comes from your mouth. I do not wish to be told either by him or you. There is no past that can be raked up? Ah, I need not have asked you that. You would have turned from him with loathing if there had been that. For the rest I am satisfied. He has artistic tastes of which I have no knowledge and with which no sympathy. He is honourable, and of a great name, he is liked, respected; he is a man whom I would have chosen myself for you, and he has the interests and welfare of the Church close to his heart——"

He stopped suddenly, arrested by the sudden whiteness of her face.

"Or what?" he asked.

"He is not even a Christian," said Helen simply.

Mr. Challoner did not reply at once. The habit of tidiness in him, unconsciously asserting itself, led him to put square the case of cards which his sister had used for her Patience. Then he turned down with his foot the corner of the hearth-rug, which Helen's dress, as she walked to the fireplace, had disarranged. Indeed, it had distressed him for some time; it was easy to trip on it. Then he spoke.

"And did you know that when you promised to be

his wife?" he asked with a scrupulous desire to be absolutely fair.

"Yes," said she.

"Then what are his religious opinions?" asked he, still scrupulous. "Does he believe in God?"

"No."

"And you knew that all along?"

"I knew it on the day when I think I began to love him," she said.

A sudden superficial flow of bitterness, just as a light breeze will ruffle the surface of some huge wave, passed over her father.

"For that reason?" he asked.

Helen looked at him in amazement.

"I did not know you could have asked me that," she said.

"And I too have much to learn about my children," said he.

Helen's eyes flashed back at him. She was afraid no longer. The talk she had had with Frank on that memorable Sunday afternoon she had put away like stored provisions; often since it had been food to her thoughts, and it was now all eaten, digested, assimilated. The instinct of individualism had no doubt often been present to her mind before, but what he said then had made it blossom and fructify. He had said, in fact, perhaps no more than she had known, though without knowing she knew it; his words had been a taper to a gas-jet already turned on. Without the taper it might have continued to escape; the taper made flame of it. And in the light of it the figure "father" was shown her as a man only, capable of using one vote, in opposition it might be to her own; but however dear and intimate he was to her, and in spite of her parentage, education and upbringing, he was still only somebody, not herself. And she, Helen, had to be herself.

"Yes, you are learning that they are people," she said, in answer to his bitterness. "Martin and I are people. I must think for myself and feel for myself. Yes, I knew that Frank is what he is, an atheist. And I love him."

Mr. Challoner looked at her a moment with terrible alien eyes, meeting her full gaze. Then he turned and went towards the door.

Instantly the daughter in her awoke.

"Father!" she cried, holding out her hands to him. "Father!"

But he passed out without turning, and she heard the door of his study opposite close behind him, and the click of a lock.

The finality, the sharpness of that click of well-oiled wards brought home to the girl even more than the bitter and burning words which had been said, what had happened, the unbridgeable breach that had opened between herself and her father. For even now, distraught as she was with the agitation of the scene, so that she felt almost physically sick, she knew that she had acted in compulsory obedience to an instinct which was irresistible; she could not call back into her own control the love she had given. Whatever else beckoned, that to her was the strongest call. And equally well known to her was the instinct in obedience to which her father had acted. Dear as his children were to him, there was something infinitely dearer, that which from the tower of the church had pointed upwards into the clear sunset sky. No assertion of individualism made its voice heard there; the one immutable Love claimed all allegiances.

Infinitely shocked and distressed as he was, Mr. Challoner did not suffer during the next half hour nearly as keenly as Helen, for the idea that she would not eventually—after pain and struggle, no doubt—see as he saw, never entered his mind. Indeed, after a few minutes the emotion predominant in him was pity for her at the necessity of the rejection of the human love offered to, and accepted by her. She would be led to the light—not for a moment did he doubt that—and the suffering would ennoble and not embitter her. Then out of pity for her, compunction at what he had done rose within him. Again he had been harsh and peremptory; not even the sacred cause he championed could justify that, nor excuse his lack of gentleness. He had left her in anger, anger as he now acknowledged to himself partly personal in its origin. So before half an hour was passed he unlocked his door, and going upstairs to her bedroom, tapped softly.

Helen had had no more thought of going to bed than he, and she let him in at once.

"We did not say good night, Helen," he said. "We were both——"

She raised her eyes to him.

"And don't let us discuss it any more to-night," she said.

"No, dear. I only wanted to say good night to you, to—to say that I am sorry for leaving you in the manner I did. You look very tired. Will you not go to bed?"

"Yes, soon perhaps."

She kissed him, and stood silent a moment fingering the lappel of his coat.

"If we did not care for each other it would be easier," she said. "Poor father! Good night, dear. Thank you for coming."

It had been arranged that Frank should bring the motor over again next morning, and drive Helen back to Fareham to lunch with Lady Sunningdale; and he made his appearance rather sooner than was expected, having driven, as he acknowledged, a little over the regulation two miles an hour. Helen had heard the approach of wheels and met him at the door. One glance at her face was enough to tell him that something—and what that was he easily guessed—had happened.

"Father is in," she said; "he waited in on purpose to see you. Yes, he knows."

"You told him? Well?"

"He said very little, but enough. Oh, Frank, it is very dreadful. He is my father. But all I said to you holds. He—you; that is what hurts so. It was awful telling him, too. But I had to."

"My darling, why?" he asked. "You should have left it to me."

Her eyes brightened.

"Ah, that was one of the reasons why I didn't," she said.

"Oh, Helen! But you look tired, knocked up."

"That doesn't much matter," she said. "Go to see him now, dear. You will find me on the lawn when you have finished. And remember it all holds. It was never shaken, not for a moment, even last night. And he came to say good night to me afterwards, poor, dear father! I have always envied him for his strength till now, but now it is just that which will make him suffer so horribly."

Frank felt in his coat pocket and took a note out of it.

"From Lady Sunningdale," he said. "She is delighted, and is telling everybody how she managed and contrived it all from the beginning."

Helen took the note.

"Go now, Frank," she said. "I can think of nothing till this is over."

She strolled out onto the lawn again, and sat down in the warm shade of the box-hedge to read Lady Sunningdale's ecstatic and desultory raptures. The scene the evening before, followed by a very restless night full of half-conscious sleep and wide-eyed awakenings, had so tired her, that weariness had brought a sort of healing of its own, dulling the keenest edge of her capacity for suffering. Breakfast had been a meal of ghastly silence, broken only by noises of knives and forks loud in the stillness. Her father had only addressed her directly once, and that to say that he wished to see Lord Yorkshire when he arrived. Breakfast over, she had written to Martin to tell him all that had happened: then Frank had come.

All sorts of awful impossible situations flapped like horrible bats about her as she waited. She pictured her father insulting her lover; she pictured Frank, stung by some intolerable taunt, striking him; she pictured with dreadful vividness a hundred things that could not possibly be. All round her hummed the myriad noises of the summer noon, and the myriad scents of the flower-garden, where still the industrious sweet-peas were prolific, mingled and were wafted in web of fragrant smell round her. It was a day of high festival in sound and smell and light and colour, a day of a brilliance that had again and again been sufficient to make her half crazy with the pure joy of living and the sight of joyous life so abundantly manifested. But this morning she was deaf and blind to the myriad-voiced noon; for in these last twenty-four hours there had come to her a happiness transcending all she had ever felt and a bitterness of sorrow, marching side by side and inextricably mingled with it, that was as immeasurably more poignant than any she had ever known, as her joy transcended all the other joys of her very happy years. Whatever might happen, life could never again be enjoyed by her with the *insouciance* of girlhood; some finger had touched her as she smiled and dreamed in her twenty years of sleep, and had awakened her. And a voice had said: "Wake, you are a woman; you shall love and suffer." Yet even now, while she shrank and winced under the pain, some secret fibre of her being welcomed it. She—her essential self—was the richer for it; life at last had touched her, sad, bitter, imperfect, but admirable life. Like a plant she had been moved suddenly out of the warm shelter of a greenhouse; hereafter the sun might scorch her, the wind tear her, the frost wither her, the rain lash her, but she was to know what it was to be rooted in the great earth, to grow, with no shelter in between, upwards toward the heavens.

All this was certainly happening to her, but as yet she guessed but a small part of it. All that her reverie, when she had read Lady Sunningdale's letter, told her, was that she was acutely unhappy because her father would suffer, and in some tremulous aerial way, happy beyond all that she had ever guessed to be possible, because she loved and was loved. The two feelings were inextricably intertwined; neither, as she knew them, could have existence without the other. And out of this tangled thicket of rose and thorn there emerged this new self of hers, in no selfish or egoistic mood, but very conscious, very vital, bleeding from the thorns, but breathing the inimitable odour of the roses.

A maid-servant with a message from the Vicar aroused her. Would she please to come into his study for a moment? She got up with a vague, dreadful sense that this had all happened before, but she could not remember the outcome; and as she walked across the lawn, the terrible, impossible pictures again flashed through her head, like scenes of a magic-lantern staring out of blackness.

The aroma of tobacco, as she opened the study door, gave her a sudden shallow thrill of comfort. But this was scarcely endorsed by the next impression. Mr. Challoner, always courteous, had no doubt suggested one of his excellent cigars, and Frank had accepted it. But the good fellowship tacitly implied by the act was here omitted. The Vicar stood with his back to the fireplace, flinty-faced; Frank sat in a big chair drawn close to the writing-table, the chair in which, times without number, Helen and Martin had sat together looking at Bible pictures after tea on Sunday. All the furniture of the study, the aromatic smell of leather-bindings that hung there, the uncompromising tidiness of it, its orderly severity, the pictures of the Roman forum, the glass paper-weight on the table, brought a sudden rush of associations into the girl's mind now that she saw Frank there too; they were all so closely knit into the fabric of her life, so intimately suggestive of that stern, tall figure by the fireplace. And somewhere far away back in her brain, her own voice in a little childish pipe whispered to Martin, "Papa's cross about something. Is it you or me?"

She took a seat in silence, and the silence lengthened ominously. Frank was looking at her with a quiet, level gaze, full of love and full of pity, and she turned her eyes away, fearing that she would scream with tears or laughter if she allowed herself to look at him. And the voice that broke the silence was quiet and level also; the whole thing was deplorably well-bred. Insults, violence, all that she had pictured to herself would have been a relief, a safety-valve for the bursting pressure that she knew existed beneath. But as yet there was none.

"I have sent for you, Helen," said her father, "to choose."

He paused a moment.

"Lord Yorkshire is on the one side," he said, "and I am on the other. We have settled it so."

"This is not quite fairly stated," said Frank, in the tone a man might use if he demurred to some argument in a discussion in which he was not really interested.

Mr. Challoner's face grew a shade paler.

"Did you say 'fairly'?" he asked.

The deadly quietness of this suddenly frightened the girl. That was a tone in his voice she knew and dreaded.

"Father!" she said, "Father!"

They neither of them took any notice of her, and Frank answered in the same gentle, objecting manner:

"You say 'we settled it'," he said. "I had nothing to do with it. You merely told me what you were going to do. That is why I used the word 'fairly'."

Mr. Challoner considered this for a moment.

"I see your point," he said. "That is so."

Then he turned to Helen.

"So, choose," he said. "I settled it so."

Helen looked at Frank a moment and stood up, love streaming round her in triumphant flood, bearing her away.

"I have chosen," she said". You know it."

Then, even in that moment when she felt so strong, when her love was to her like a draught of wine, or meat to the hungry, her strength utterly failed her, and she buried her head on the cushions of the sofa where she had been sitting and burst into hopeless, hysterical sobbing. She was not capable of more; all had given way, and she lay helpless, sobbing, sobbing, as if to sob her heart out.

But four hands were busy about her, and as the stress of her seizure began to leave her, she heard two voices, for the moment one. And one said, "Helen, darling," and the other, "Helen, dear;" and one said, "If you would be so kind, Lord Yorkshire; there is some water on the table;" and the other said, "Helen, would you like to drink a little water?"

For two men, in nature, in sympathy, in religion, poles apart, were bound together for a moment in the necessity divine and human of comforting the weak, of giving help to a sufferer. She who suffered was loved by them both; and though the distance of fifty poles could not span the difference between their ways of love, that was sufficient.

For myriad are the ways of approaching the throne where all love dwells; from East and West and North and South, those myriad ways converge and meet. But at present East and West, being human, and thinking that they were going in opposite ways, could not foretell the meeting. But the Centre knew.

By degrees she came to herself again, and one said, "Some other time," and the other, "Not again now, Helen." So of the three she was the only one who was resolved to go on, to have this ghastly spiritual surgery finished. Though she had chosen, she knew there was more that had to be said.

She cast one glance at her father, but, her physical weakness over, his pity, she saw, was over also. A gulf, immeasurable by leagues, had opened between them, and though not even yet did he despair that they would be ever united, it was she who must come to him. From the firm rock on which he stood, he knew, so he believed, that he would never stir a step.

She pushed back her hair from her forehead.

"I don't know why I did that," she said. "It was stupid of me. Give me a minute."

She got up, still a little unsteadily, and played with the pens in the tray on the writing table, recovering herself. Then she turned suddenly to her father.

"Father," she said, "you can't mean what you say. How can I choose between you? What are you asking me to do? What do you mean?"

"I mean exactly what I say," he answered, with the same dreadful quietness. That which had not seemed possible to him last night, that she would really choose as she had chosen, had become more than possible. "You choose between us. Are there words in which I can make that clearer? If you choose me, you say good-bye to Lord Yorkshire here and now. If you choose him, you are to understand that you cease to be my daughter. I will not be at your wedding. I will not see you afterwards. You shall not be married from this house, nor if I could help it should you be married in this church."

Then suddenly the quietness of the scene was shattered. As if by a sudden flash of lightning, all that Helen's choice implied, her rejection not of him alone, but her rejection of all in the world that he held sacred, was made dazzlingly clear to him. At that his self-control gave way, and as his voice rose louder and louder, he beat with his clenched hand on the edge of the marble chimney-piece, so that the knuckles bled.

"Understand what you are doing," he said, "and let me tell you, so that there can be no mistake. You will promise to love, honour and obey an atheist, an infidel, one who denies God and His Christ. You will have to say you do this according to God's holy ordinance. That from you, in church, Helen, and a lie! It cannot be by His ordinance, for by your act you turn your back on the faith that has been yours from childhood till now, on all you have believed to be sacred. And what of the end? What of the life to which this is but a prelude? What of him, your husband, then? He that believeth not shall be damned. I would—I would sooner see you in your coffin than standing by the altar with this man. I would sooner see you his mistress——"

His passion, springing though it did entire from his own intense and fervent Christianity, had suddenly shot out into a bitter and poisonous blossom, and as that flared through the room, he paused a moment and looked at her, as she stood before him in the beautiful whiteness of her girlhood. Her physical weakness had altogether passed, and, except that she took one step back from him in involuntary disgust and shrinking, you would have said she was listening with quiet, incredulous wonder to some tale that did not concern her. But as he paused, hardly yet knowing what he had said, knowing in fact only that no words could be strong enough to express the intensity of his conviction, she turned from him.

"Come, Frank," she said; "let us go."

Frank also had risen with a sudden flush on his face at those intolerable words, an answer springing to his lips, and moved quickly towards her with some instinct of protecting her. But her tone checked him, and he followed her to the door. She had already opened it, without further speech or looking back, when her father's voice, scarcely audible, and broken and trembling, stopped her.

"Helen," he said. "Indeed I did not think or know what I said. But, my dearest, what are you doing? What are you doing? For Christ's sake, Helen, who died for you."

Frank had passed out: whatever more took place between them was not for him to hear. Then the door closed behind him, leaving father and daughter alone.

"For Christ's sake, Helen," he said again.

She came back to the hearthrug where he stood.

"Oh, father," she said, and paused. That was all the reproach he was ever to hear from her. "You are making it very hard for me."

"Yes, I am making it as hard as I can. I am bound by my duty to God to do that. If I knew how to make it harder I would."

"You cannot: you have said all that can be said. And I have nothing more to say. Let me go now."

She kissed him gently.

"Poor father!" she said, and left him.

Mr. Challoner stood long where he was when she had gone. Never before perhaps in his whole life had another will come so actively and stubbornly into collision with his, and never before certainly had he felt so overwhelming a sense of spiritual desolation. Eager and strenuous all through, it was in the truths of the Christian faith that he found the incentive of his life, from it sprang all the earnestness and deep sense of duty in the man, to it was every effort and deed of his dedicated.

"But what have I done," he half moaned to himself, "that this should come to my house, and to one for whose faith and upbringing I have to answer? O Lord, if it is through any fault of mine, let me learn for what deadly sin this punishment is sent!"

Indeed he had spoken no more than the truth, bitter and brutal though the truth was, when he told Helen that he would rather have seen her in her coffin than by the altar with her lover. And now he took no account of his personal sorrow; the yearning that she should accept her father's wish and guidance as such was nonexistent in him, killed by the stronger motive. All his personal relations with her of trust and affection, which, to the best of his power, he had built up for years, were voiceless now: simply he strove for a soul—and that dear to him—in danger imminent and awful. The rigid Puritan note was here, and he would sooner have mated her with a thief or an adulterer, since such might repent and be saved, than with a reasoned atheist.

Then, in a horror of great darkness, he questioned his own spirit. "How had he failed?" and again, "How had he failed?" Never had precious plant been more hedged about from frost or untimely blighting of March winds than had his daughter been folded from all that could conceivably have stunted or weakened the one true growth. From the time when her lips were wet with a mother's milk, God's counsels, verse by verse and line by line, had been the guides and counsellors of her life. What had he left undone that he could have done? Had any remissness of his own hindered growth where it should have helped? He searched the years for his fault, but among all his failures and weaknesses and harshnesses, he could not find that even for a day had he let anything else take precedence of the greatest and the only thing in the world.

And now at the end she would mate with an infidel—a man, according to his idea, whose intimacy was more to be shunned than that of a leper, or of one who was tainted with some deadly and contagious disease. That, at any rate, would only kill the body, but Helen had chosen as the friend and companion of her nights and days one whose soul was sick with a more fatal disease, the end of which, ordained and appointed of God, was eternal death. It was too hideous to be credible, it was too hideous to be conceivably just. And the fact that he could think that gives the measure of his soul's anguish.

God sets a limit to human misery; for it happens that the tortured brain, tired with suffering, lapses into a state of semi-sensibility; or again, since one cannot feel pain on account of another, unless the other is dear—the pain felt varying indeed in proportion to the affection felt—the joy of love is always mingled with it. It was so now with Mr. Challoner. Had he not been Helen's father, had he not loved her, he would have cared less. But she was his daughter, his own girl, whose sweetness had all her life made sunshine in his home. He had said an intolerable thing to her, and for reproach she had still given him gentleness. In the keenness of his own suffering he had forgotten her; he had forgotten even, except for that moment when she had broken down, that she must be suffering. So he went out after her.

She was standing at the door with her lover, and he went straight up to them. Even the sight of Frank there gave him no pause.

"It has been a dreadful morning for us all," he said, "and selfishly I had forgotten that others besides myself were unhappy. God knows what is in store for us, but we can do no good by being bitter, as I have been. Let us —yes, you too, Lord Yorkshire—let us all join hands a moment. We are His children, are we not? We——"

His mouth quivered, no more words would come, and

they stood there a moment, all three, hands clasped Then, feeling that his self-control was utterly giving way, he left them, and went back to his empty room.

CHAPTER VIII.

Helen was sitting on a pile of crimson cushions in the stern of a Canadian canoe, while from the middle of the boat Martin, with shirt-sleeves rolled up over his brown elbows, paddled her gently along the reaches of the upper river at Cambridge. The dryness and heat of this glorious summer had made the river very low in places, and his feet also were bare with flannel trousers rolled up to the knee, for again and again he had to get out to pull the boat round snags or over shoals, where the depth did not allow it to pass with the draught of two passengers. To the right, across a stretch of meadow stained brown with length of summer suns, rose the tower of Grantchester church, embowered in trees, and the booming of the mill sounded drowsily through the still air. Close to the river, however, a vivider tone of colour prevailed, tresses of water-side foliage dabbled in the stream, and tall, slender trees made a shelter from the heat, where cows, a classical example (and so not appealing to Martin, who splashed water at them) of unbustling life, chewed the cud, and looked with large incurious eyes at the gliding constellation of the Twins. Between them in the boat were packages containing lunch, for Martin had taken a complete day off his studies in recalcitrant languages, and was devoting himself to Helen, who was staying with an aunt, Lady Susan Arne. Dr. Arne, her husband, was tutor at

King's, at which seat of learning Martin pursued his antipathetic labours, and had the reputation of being the greatest authority living on the metres of Greek choruses.

Helen had left Chartries a couple of days after the crisis of her love affair, at the suggestion of her uncle, to whom she had confided it.

"I will walk back with you to the Vicarage, Helen," he had said, "and persuade your father, in case he needs persuasion, to let you go away at once. Your being with him just now only keeps the wound open. Go away; it will heal better so. Just now, after that scene, you can only torture each other by your remaining there. Poor dear child!"

"Yes, but 'poor father,' too," said Helen.

"Certainly. Come to Chartries, if you like."

Helen took his arm.

"That is so good of you, Uncle Rupert," she said; "but I think I should like to go quite away if father will let me. I think I should like to go to Cambridge. Martin is there. And Martin is *so* good for one, if one is—well, not very happy."

"Yes, that's a good plan; you can stay with Susan. My dear, I'm more sorry for you than I can tell you, and also I am sorry for your father. You and I both know him, and we both love him, and though we are made very differently, we know how—how splendid he is. And how big."

"I know," said she. "I feel that if I could only persuade myself he was narrow, I should care less. But his huge single-hearted devotion to—to God cannot possibly be called narrow."

They walked on in silence a little.

"But that is all I can do for you, Helen," said he. "Nobody can really help you except yourself; we can only alleviate things a bit. You have made your choice, absolutely, I gather?"

"Am I being a selfish, egotistic little brute, Uncle Rupert?" she asked.

"Not according to my view, which is that when a thing concerns you so intimately and vitally as this, it is nobody else's business. Not even your father's," he added.

A good deal of persuasion, as Lord Flintshire found, was needed. At first his brother would not hear of Helen's going, for he said that her departure was shirking the situation. What made him yield was the suggestion that the situation, if not shirked, might make her really ill. And a hurried interchange of telegrams led to her arrival at Cambridge the next evening.

The expedition to-day had started rather silently, and Martin decided that as Helen did not at present want to talk about her affairs, the best thing to do was to be completely futile, foolish and garrulous. For years he and Helen had adopted this method of treating each other's depression, and it was sufficient for one to say, "Hump. Play the fool," for the other to understand that until further notice he had to talk rot. This was a device, by the way, which neither had ever employed when Mr. Challoner was in a similar mood. He would probably not have understood it.

Martin stood up in the boat which had stuck, and peered into the water.

"The great thing," he remarked, "as the White Knight said, is to guide against the bites of sharks. He had steel anklets. Ow! why do they take the sharpest stones in the world and place them where I want to step? I'm bleeding like a pig."

He stood precariously on the other foot, and examined the injury.

"A pig," he remarked fatuously, "that has not yet had its throat cut. Helen, how fat you must be getting! You weigh tons. We'll have to throw the lunch overboard. Or perhaps it would be simpler if you stepped ashore for a moment. You can easily step onto the bank from there."

He pulled the canoe over the shoal and took it where she could get in again. She laid her hand on his shoulder as she stepped in.

"You darling!" she said. "You can stop now. I'm better."

"That's good work," said Martin. "Because really I was beginning to run rather dry. You mightn't have thought it."

"I didn't. I had no idea of it. I thought there was any amount more."

"I can manage ten minutes more, if you like," said Martin.

"No, I'm going to talk now. Martin, if you look suddenly grave like that, I shall begin to laugh."

"Well, give me a couple of minutes," said the outraged Martin. "We always have an interval after the rot before we begin to talk. Otherwise, you know we always laugh. One always laughs at anything abrupt. Don't you know the story of the man who was suddenly told his wife was dead? Just like that. He said, 'Oh, how shocking!' and burst into shrieks of laughter. And he was really devoted to her, and never smiled again for years."

Helen gave up all attempts at gravity, and the two foolish Twins laughed till they were completely exhausted, while the Canadian canoe went slowly circling round and round down the river.

So they landed and lunched, as Martin refused to drag the boat any more till he had eaten, and by degrees recovered themselves. Then, taking to the canoe again, they paddled and talked.

"It has been dreadful at home, Martin," said she. "Father hardly speaks at all. He has been very gentle since that scene with Frank and me, yet even that was hardly so bad as his silence and quietness now. He is suffering horribly too: I am sure of it. Sometimes I see him looking at me with a sort of appeal in his eyes like a dumb animal. That is the worst of all: I feel such a brute."

"You suffer too," said Martin quickly.

"I know, but though they all, Uncle Rupert, Lady Sunningdale, think I am right, that doesn't make me feel less of a brute. Besides, there is no 'right' about it. I can't give him up, and father can't bear it. And every evening he uses the prayer for Jews, Turks and infidels."

Martin frowned.

"That is not good manners," he said, "with you there."

"Oh, Martin, manners don't come into it. The truth of father's beliefs is so overwhelmingly real to him that he can't think of anything else. That light is so strong that he can see nothing but it. It is soberly the whole world to him."

"But it isn't as if Frank was immoral," said Martin.

"I believe he would mind that less," said she.

Martin swung the canoe round a half-submerged tree-trunk, where the water sucked and gurgled.

"But how unreasonable!" he cried. "Frank can't help his want of belief. But we can all in some degree help making brutes of ourselves."

Helen sat up suddenly, causing the boat to rock.

"I can't live my life on other people's lines," she said, "any more than I expect others to live theirs on my lines. 'I am I.' I remember Frank quoting that to me the Sunday he walked back with me from Chartries. That has been like leaven; it has fermented and expanded within me. But, after all, is it only another way of saying, 'I shall be as selfish as I please?'"

"Of course not. That is what people think who haven't got any individuality of their own. Lots of people haven't. They are like mirrors, slightly cracked, which reflect with certain dimnesses and distortions what is put opposite them. They say individuality is selfishness. What bosh!"

"Aunt Susan hasn't got any," remarked Helen, letting the conversation drift away a little. "It is that which makes her so restful. Her mind is like a cushion. It is quite soft, and if you lean on it you make great dents in it."

Martin remained quite serious, staring at the water with vacant black eyes.

"Poor father!" he said at length. "Just think; you and me, Helen. He must find us awfully trying."

"I know; and he continues to love us so. It is that which makes it so dreadful. Oh, Martin, do get through your stupid examination. Do turn out satisfactory, as I've been so eminently the reverse. It is your turn."

Martin transferred his gaze to his sister.

"Don't think there's much chance of it," he said.

"Of your getting through?"

"I might manage that. But there are other things. The career I propose, for instance."

"But he's reconciled to that," said Helen. "That's nothing new."

Martin paddled on without answering this, and Helen looked at him rather closely.

"There is something more," she said. "What is it? Is there not something more?"

He brought the boat up to the bank in Byron's pool, where they were to disembark.

"Yes, there is," he said. "At least there may be. There is no use in my telling you now. If it happens, if I am sure it is going to happen, I will tell you beforehand. I promise you that. And now I think we won't talk any more about it."

But a sudden uneasiness seized the girl.

"Promise me one thing," she said. "Promise me it is nothing disgraceful."

Martin looked rather injured.

"No, I have not been stealing hens," he said. "And it is compatible with the highest character."

Helen looked at him a moment in silence.

"Then I'm not afraid," she said. "And I will try not to guess at it until you tell me."

The afternoon was intensely hot, and having arrived here, they settled that a boat under trees was far more to the point than walking under the blaze of the sun, and Helen merely reclined more recumbently on a pile of cushions.

"I think we will go for a walk to-morrow, Martin," she said, "instead of to-day."

"That may be. By the way, I met last week that nice girl who was down at Chartries on the Sunday when I got into so many rows. What was her name?"

"I haven't the slightest idea," said Helen.

"Yes, you have. Oh, I know—Miss Pl—— Oh, yes, Stella Plympton."

Helen did not answer for a moment.

"Well, I shall go to sleep," she said. "Martin!"

"Well?"

"You did that remarkably badly," she said. "A cow could give you points in dissimulation. You remembered her name perfectly."

Dr. Arne, at whose house on the Trumpington Road Helen was staying, was probably as nearly happy as is possible to the sons of men who have so marked a genius for discontent. Whether his happiness was worth much, and what it all came to, is another question; but happy he was, an affair of immense importance not only to himself, but to all on whom his imperturbable serenity shone. For Providence had endowed him with an apparently insatiable curiosity about the chorus-metres in Greek plays, and also with an intuitive perception as regards this extremely difficult and no doubt fascinating branch of knowledge, which had proved itself capable of being trained into something approaching the perfection of acumen. His intellectual ambitions were thus completely satisfied, and being without any passion but this, which the fact that he was tutor of his college enabled him to gratify without stint, there was really no possible chink at which the bitter wind of discontent could enter and make draughts. The same good fortune had attended his marriage, for he had wooed and won a woman of good birth and breeding, whose only desire, as far as he was aware, was to make her husband, not happy—he was that already—but comfortable. Extremely edible meals were offered to his notice at hours of his own choosing, no sacrilegious hand ever disturbed the papers in his study, his wife walked with him after lunch, and, unless they had people dining with them or they were themselves bidden to other feasts, played picquet with him after dinner. His mode of progression along roads was naturally a little quicker than hers, his play of the hand at cards a shade less mediocre, and in consequence he lived in an atmosphere of slight domestic superiority. The same atmosphere, though not domestic, surrounded him in his studies; for, to make a rough statement of the matter, he knew rather more about Greek chorus-metres than anybody else had ever done. His bodily health, moreover, if not exuberant—he would have found exuberance very trying—was excellent; he appeared, in fact, to be as immune to the frailties and disorders of the flesh as he was to any unsatisfied cravings of the spirit. He was also childless, and though he was not consciously grateful for this, he was aware that he desired neither more distractions, anxieties, or even joys than he possessed in such completeness.

Lady Susan Arne had been compared by her niece to a cushion, and indeed the superficial similarity—not indeed in point of looks, for Lady Susan was remarkably well-favoured and inclining to thinness—in the nature of the two was extremely striking when once it had been pointed out. It was true that if one leaned on Lady Susan's mind there was no firm resistance; only a large dent seemed to have been made in hers. But Helen, with a certain impatience in her survey, had overlooked the existence of a permanent dent there, a thing entirely foreign to cushions. She, Helen, it is true, might lean and make a dent, and the next person who, so to speak, shook Aunt Susan up, or leaned upon her in another place, would (still in Helen's view) efface the first dent; but in a corner of her where no one ever thought of leaning or looking, there was a permanent and uneffaceable dent. This was made in the first place by the ungratified yearning for a child of her own; it was now daily renewed by the knowledge of its impossibility. There was in her, in fact, a potential vitality, which under other circumstances might have made of her a woman, not a housekeeper, and have given her points more directly in contact with life than were picquet and constitutionals. As it was she had experienced none of the divine unsatisfiedness which fulness of life alone brings with it; she knew only the content of a rather empty existence. And Helen, judging with the impatience of youth, which is akin to the impatience of kittens or puppies with inanimate objects that will not come and play with them, had overlooked this. For in truth Aunt Susan was not inanimate: tucked away in a corner of the cushion was a real live thing that groped for life and light, and she, the apparent individual, was like a room made ready for the reception of guests, chairs and tables in order, games put out for their entertainment, but until the guests began to arrive the room was in darkness; Aunt Susan stood there, matchbox in hand, so to speak, waiting for the first ring at the bell to light up her tapers, and show how orderly, how fragrant, how charming (a little old-fashioned too) her room was, how thoughtfully arranged for the pleasure of others. But no ring had yet come at her door bell, and she still stood there, very patient and still smiling, but still waiting.

Lady Susan, on Helen's arrival, knew only vaguely that something uncomfortable had happened at the Vicarage, but Helen the first evening she was there had confided to her, rather as one may confide on cold nights to one's pillow or to bed-clothes tucked round the neck, the history of the last few days. But she neither knew nor would have guessed it possible that the news had kept Aunt Susan awake half the night, and that while she herself was up the river with Martin, her aunt had gone about her household businesses, and taken her walk with her husband, in such a tremor of excitement that he had to hurry after her, instead of hanging on his step to wait for her. In all these tranquil years at Cambridge she had never been brought into contact with a thing that moved her like this. The gentle ministrations in which her years were passed had not touched her emotions, which, had not her yearnings for a child kept them alive, would probably long ago have fossilised. But those yearnings had nourished and rendered mature their sweet, delicate sensitiveness; and now when they were aroused, though even in this second-hand manner, they responded instantly, gently vibrating, not with a crackle of dry autumn leaves, but like foliage of aspen in the breath of Spring.

Helen got back to this house of quiet towards five in the afternoon, and found her aunt and Dr. Arne at tea on the lawn behind the house. The latter, however, soon went indoors to enjoy—literally enjoy—his couple of hours' work before dinner, after forewarning them as to possible dampness on the grass after sunset.

"And have you enjoyed yourself, dear," asked Aunt Susan pleasantly, "and was the lunch I gave you really sufficient? Dear Martin has always such a beautiful appetite. It is a pleasure to see him eat his dinner."

"Yes, dear aunt, we had heaps. And it was all so good, and so beautifully done up. Exactly like you."

Aunt Susan, who always looked like a kind little animated Dresden shepherdess, flushed a little.

"And so you had a nice day?" she said. "And no

upsets? Martin is so reckless on the water. Dear Helen, is it quite wise to take off your hat? It may turn suddenly chilly."

Helen laughed, and threw it on the grass.

"No, no upsets, and quite wise, Aunt Susan. But a nice day? There was everything to make it nice externally, but one's nice days are made inside one, I think. And just now my machine for making nice days creaks and groans; it is out of order."

Aunt Susan, though far too shy to take the initiative, was longing for the least thing that could be considered an introduction of this topic.

"Do you know, dear, I lay awake half the night thinking of you and your trouble," she said.

"O, I'm so sorry!" cried Helen. "I ought not to have told you so late last night. Selfish little pig I am!"

Aunt Susan patted her hand gently.

"Dear, it was delicious," she said, "lying awake and thinking about you. I am afraid I actually enjoyed it. Not that I am not very, very sorry for you and your father and Lord Yorkshire; but when I said it was delicious, I meant it was so real, so alive, so very interesting. I don't think I have lain awake more than a few minutes in the last couple of years, and that was when your uncle had the influenza. And then it was only his cough that kept me awake; I was not anxious, for he had it very slightly. Now, if you do not mind talking about it, do tell me more. You told me just the facts. Tell me what you feel: how does it touch—I am so stupid at saying things—not what you will *do* only, your actions, but yourself?"

The question implied a perception with which Helen had not credited her aunt.

"Ah, what a difference there is between them!" she said quickly. "One's actions may so frightfully belie one. What one does is so often a parody of one's best. One's worst part acts, while one's best does nothing, turns its face to the wall like Hezekiah. Or, or—" she was still kindly trying to explain to this dear little Dresden shepherdess—"one's actions are often like an unsympathetic repetition of something one has really said, which gives quite a different meaning to it. Do you understand?" she asked eagerly.

"Yes, dear, quite," said Lady Susan. "Surely everybody understands that. All the same it is our business, if we are kind and good at all, not to be harsh or hard in what we do."

Suddenly Helen's eyes were opened. In a flash she saw that she had been doing what she deprecated, and hitherto had judged Aunt Susan merely by her actions. With the impatience that was so very characteristic of her, she had observed her ordering dinner, taking the walk, playing picquet, and otherwise having a great deal of rather fragrant leisure with which she did nothing: from this she had drawn the conclusion that there was, so to speak, no one really there, only a punctual little domestic automaton. She had been so taken up with the fact that others did not understand her, did not allow for her individuality, that she had as yet never taken the trouble to consider whether these others also had not their own individuality, equally to be respected. Aunt Susan, she would have said off-hand, had none, yet here she was referring to as a mere commonplace what was still to Helen a blinding discovery. And she went on talking with a freedom and a certainty of being understood, that she associated only with the beloved twin. . . . .

"Well, it is just that," she said. "Anyone, you, Uncle David, anyone may say it is merely heartless, merely selfish of me to go my own way, to pay no attention to the wish—ah, it is much stronger than that—of my father. Or you may think that I don't really know how strong his objection to my marriage is. I do know, I fully know. And knowing that, knowing also that he is my father, that I owe nearly everything to him, that he loves me, and I love him, I am going to do, you may say, *as I choose,* throwing away all the love and the care he has spent on me, repudiating my debts to him. But I don't. Oh, Aunt Susan, I don't throw away his love, or repudiate my debts. It is not fair to say that. Simply I can't help it—I must. Something has come which is stronger than everything else. Ah, Aunt Susan, you know what it is."

Lady Susan's delicate little china-looking face flushed suddenly.

"Yes, dear, I know," she said. "At least I know some of it. We women are meant to be wives and mothers. I know half of what a woman longs to know. And the half I know, dear Helen, is so very fine that it is worth making some little sacrifice for it."

"Sacrifice?" asked the girl.

"Yes. I cannot tell you in great language what I mean, because I am not great in any way, so I will give you my advice in one short word. Wait. Love is so good that it will not spoil by being kept, it will only get more mature, more exquisite. And in the meantime you will have proved yourself a good daughter too."

"But why—why?" asked the girl. "Nothing will ever change what father feels about it, nor what I feel. It only means that for six months more, or for a year more, or however long I wait, he and I will go through those dreadful days. It is awful at home, Aunt Susan: you have no idea how awful. If it would get any better with waiting, I would do as you suggest."

The older woman was still smiling in the habitual way which Helen had so often thought so meaningless, so objectless. But now as she looked she saw there was a very cheerful patience about the smile, which somehow she had not noticed before.

"It is true it may not get better with waiting," she said. "But you will have done your best, not only thought your best. You will have made your action not, as you say, the parody of yourself, but the faithful expression of your very best self. You will have put your speech into no unsympathetic mouth, but into the mouth of a fine actor."

Another current seized the girl, sweeping her impetuously away. She laid her hand on her aunt's knee.

"Are you unhappy, Aunt Susan?" she asked. "Oh, I hope not. I always thought you were so contented, so—so occupied with all the duties you do so well."

Lady Susan, with the only movement of impatience that she had made perhaps for years, swept her hand away.

"Ah, that is because you are young," she said, "and because you think that anyone who feels an impulse must act on it, if she wants to realise her life. It is not so. You know what I have always called you and Martin, the Volcanoes—dear Volcanoes. When you feel pressure you burst, and scatter burning ashes anywhere and everywhere, and say with great good-humour, 'But I am I; if I want to burst, I must.' And when you see an old woman like me, just getting through the day's work, day after day, week after week, with a little dinner-party here, and a little walk there, and a little ordering of the household all through, you think, 'Is that all? Is that life?' And I answer you, 'Yes, that is life.'"

Helen was silent for a moment, suddenly aware that for the time it was perhaps wiser to listen and attend than talk about her own individuality.

"Tell me, tell me," she said.

"My dear, there is very little to tell," she said. "But you in your heyday do not allow, it seems to me, for the fact of other quiet people living and feeling perhaps just as much as you do. Because you feel a thing you scream. You will learn to feel a thing, we hope, without screaming. I think young people tend to scream rather more than we used. They call it living their own lives. That possibly may be a mistaken, or anyhow a misleading name for it."

Again Helen had no reply. But this did not seem to her at all like want of individuality. There was no screaming, it is true, and no assertion, but just as certainly there was "something there." And to do her justice she respected that. But her aunt paused also, waiting for her answer, and after a minute she spoke.

"Live your own life then in talk with me," she said. "Let me understand it. It is quite true, Aunt Susan, I have judged as if there was no other view than mine, while the whole time my complaint—no, not that exactly, but you understand—has been that other people behave as if there was no other view than theirs! About you, for instance; I didn't know—I didn't guess. I thought you were—you were what you appeared."

Lady Susan seemed to repent of her hasty movement, and recaptured Helen's soft brown-skinned hand.

"Yes, dear, I am," she said quietly. "At least, I choose to let that be my outward expression of myself, the expression by which you, Martin, anybody may judge me. That certainly is my affair, and nobody else's."

She ceased stroking Helen's hand a moment and looked up at her.

"But, dear, would you like to come inside for a moment? There is only one thing there, but it fills my house. Oh, Helen, if I had had a child!"

At that all the girl's nature rose.

"Ah, dear aunt, dear aunt!" she said.

Lady Susan's pretty patient smile did not leave her lips, nor did any tear come to her eyes; the sorrow was too old and too eternally alive for her to weep over it now. And she went on quite quietly:

"If only I had been given the chance even to be made as unhappy as you are making your father, dear, I should have loved it so. But it was denied me, and by no fault of mine. So I am learning, I hope, not to grumble. Ah, but it is hard sometimes, and I think I miss the joys of love as you would count joys, Helen, less than I miss what you would count its sorrows. But those are its opportunities, dear, its possibilities in self-denial and self-abandonment. That is Love triumphant, not crowned with roses, but crowned with sharp, beloved thorns. And the tragedy of love is when there is none for whom it can sacrifice itself."

She stroked Helen's hand again gently.

"Make yourself complete, dear," she said. "There I am entirely at one with you. But remember, our souls are like rose trees, I think. You cut and prune them, if you are a wise gardener, for you know that by the cutting, the renunciation, you do not check or hinder your development, but you encourage it. You will be the more fragrant, the fuller of blossom by that which you might hastily say was a piece of cruelty, a stunting of your growth."

Her kind eyes looked away from Helen, and out over the sun-baked lawn, bordered with flower-beds, in which, clearly to comply with preconceived notions of a garden on the part of a gardener, lobelias were set in a formal row in front, and behind them terrible speckled calceolarias and hard crude geraniums. That garden had often seemed to Helen very typical of her aunt, it was orderly and completely conventional. Beyond Dr. Arne's study windows looked from the red-brick house across the grass, and from where they sat she could see him at a table littered with papers and manuscripts, with head bent over his work, or rising now and then to consult some book of reference which he took from the volume-lined walls. That sight also had often seemed to her very typical; the Cambridge professor was at his work (as indeed it was most right and proper that he should be), but that to him was all. His little life was bounded with books; on all sides stretched limitless deserts of particles and chorus-metres. But now for the first time Helen knew how erroneous all her judgments with regard to Aunt Susan had been; for a real heart beat there, and it was somebody, somebody very distinct and individual who ordered dinner and played picquet. Her life was not negative, emotionless; it was only Helen's own obtuseness of perception that had so labelled it. Instead it was sad in spite of all its quiet cheerfulness; it was as sad as the level rays of the sun striking hazily across the lawn, as sad as the grey spires of King's, which rose against the clear hot blue of the sky.

And the pathos of it suddenly moved her. Was that all that the good fairies had brought to her aunt's cradle, just to grow quietly and gently old, she who might have been so fine, missing all the joy and riot of life, missing, too, the crown of womanhood? "To live, to live!" that demand was battering at her doors with buffets that made the panels start. Yet here was the dear aunt, who had heard often the same insistent visitor, old, but sweet and unembittered, though it had never been given to her to let him in, knowing all she had missed, yet not soured at having missed it.

"Oh, Aunt Susan," she cried, forgetting herself, forgetting all else in a young creature's somewhat insolent pity for the old, "Is it not too sad? Is it not too terribly sad? Is that everybody's fate, just to get older and older?"

Then, with the strong, unconscious egotism of her years:

"And me?" she said. "Will that happen to me, too?"

"What? Sadness? Yes, dear Helen, I hope so. No woman is worth very much until she has been through a good deal of sadness, a great deal of wanting that which she cannot get. I hope you will go through that. But, dear, if you turn bitter under it, you had almost better not have lived, and certainly you had better die, for death is better than bitterness. But if you take the woe and the sadness, which is inseparable from life, without bitterness, it strengthens and cleanses you. And you will certainly emerge from it a far finer creature than if you had never been through it. Emerge? Ah! it may last to the day of your death, but what then? What does that matter?"

There was a long silence, and the shadows grew and lengthened on the grass, as Helen sat unseeing but absorbed, gazing wide-eyed in front of her. She felt ashamed, humiliated at her own blindness; she had thought of her aunt as some dweller in the valley, while she herself was climbing the snowfields far above with eager, untiring foot. But now at the summit or near it she saw sitting the quiet, patient figure, so high up that she had not seen her before.

Then in her gentle voice Aunt Susan broke in on her reverie.

"There, dear," she said, "the sun has set; let us go in. And do tell me, Helen, before you go home what you decide to do about this very difficult choice that is before you. Of course you will not give Lord Yorkshire up; I think that would be very wrong. Do not be hasty; do not judge quickly. But do confide in me again, if you can. It is a great privilege, you know, for old people to be confided in by the young. Come; it is time to dress; there are a few people to dinner. Ah! Martin comes, too; I had quite forgotten. Dear me, how careless! I must go and see if there is enough to eat."

Helen rose and gave her a great tempestuous hug.

"You dear, you dear!" she said.

And then Aunt Susan, after her excursion into realities, hurried to the kitchen, the excellent housekeeper again.

There must have been something in the conjunctivity of the Twins—except, indeed, in the "House" of the Vicarage at Chartries—which disposed the beholder to indefensible levity. London had felt their spell, and even Cambridge, it appeared, that home of sweet and sober seriousness, went a little off its head about them. The spell, whatever it was, lay in their combination. Helen alone, it is true, could rouse that impulse of social gaiety which is evoked so easily by a girl's beauty and high spirits, and Martin could make other people enjoy themselves by the sight of his own enormous power that way; but it was when they were together that resistance was clearly hopeless, and it is worthy of record that to-night, after dinner, Dr. Arne and a Professor of Poetry, with their respective wives and the Twins, "played ghosts" in the garden. Why these elderly people did it, they could not have told you; but Martin proposed "Ghosts," Helen explained it in three sentences, and the studious shades were awakened and appalled by wild shrieks.

For the night was dark and moonless, and while five out of these six foolish people hid in asparagus beds, behind tree trunks, in the wood-shed and in other black and dreadful places, the Professor of Poetry (selected by lot) was in honour bound to make the complete circuit of the garden, conscious that at any moment a ghost with curdling yells might spring out on him, or, even worse, scuttle quickly up behind him, or, perhaps worst of all, he might suddenly be conscious of a small crouching figure by his side which accompanied him in awful silence, ready to break forth into who knew what hideous and babbling speech? Thus one eye had to be kept on this dreadful object while simultaneously the whole attention had to be on the alert in case of some new *revenant* from the bushes. The Professor was a man on whom, as far as was known, the imputation of cowardice had never yet been laid, but at the first attempt to make the black circuit of the garden he found he could not possibly face the corner by the wood-shed, his nerves being already utterly unstrung by a vague form that groaned among the gooseberry bushes. He paused while still a few yards distant from this dreadful being, and then fled with flying coat-tails back to the house, where in the safety of the lit drawing-room he wiped the dews of strangling anguish from his forehead, and called lamentably on his courage.

"'Not a glimmer from the worm, in the darkness thick and hot,'" he half moaned to himself. "Oh, this will never do! I am aware it is probably only Dr. Arne, and I am not really frightened of him. Come, come!"

And with his heart in his mouth, he set out again on his fascinating and abhorred errand, murmuring over and over, "'In the darkness thick and hot. In the darkness thick and hot. In the dark—' Oh, dear me, what is that?"

The poor Professor suffered for his momentary panic, for Helen had in his hour of weakness in the drawing-room changed her place to behind a large flower-tub which had concealed nobody before. Consequently he approached it inattentively without caution or misgiving, to be confronted shuddering by a flapping form which gasped and panted.

He made a fruitless appeal.

"Dear Miss Helen," he said, "I can't go on. I really do not think it would be right. My work will suffer. But is it Dr. Arne among the gooseberry bushes, or is it Martin? I think I could run as fast as Doctor Arne, but if——"

Hoots of unearthly laughter assailed him on the other side.

Afterwards they played Dumb Crambo. Lady Susan in a college cap and a dust-coat of Martin's was Alfred letting the cakes burn; at another time Doctor Arne found himself to be Cleopatra with Helen as Mark Antony. He chose his dresses from Helen's wardrobe—they were much too large for him—with immense care, and subsequently applied a paper-weight in the form of a snake to his bosom; the Professor of Poetry became a prize-fighter; his wife, a godly and virtuous woman hitherto, unexpectedly turned out to be Peace the murderer, and did a deed of blood with immense gusto and a paper-knife. Yet all the time nobody asked himself why he did these silly things; the Twins had said it was to be so, and that was enough. At their order, too, it seemed as if the golden gates of youth had swung open, and the tired and the patient and the elderly and the wise were bidden to enter once more, and be children again.

Helen's visit to Cambridge had been restricted by no statute of limitations in regard to time; and the days passed on, the vague "few nights" growing to a week, and the week to a magnified fortnight. For these quiet uneventful hours, in which (except when the Twin was with her) even the ticking of clocks seemed muffled, had an extraordinary and growing charm for her, since she had learned that behind the outward placidity in her aunt there lay a very real inward life in which she longed without possibility of satisfaction, and suffered without bitterness. That, somehow, to the girl seemed to lift up and consecrate Aunt Susan's homely little employments, which, sweetly and patiently performed, became symbols and signs of a very beautiful character, and that which Helen had thought dull, unperceptive, unemotional, was now lit from within, as it were, by the uncomplaining cheerfulness which gave such gentle, unquestioning welcome to the limitations set about her. For Lady Susan, so her niece had now learned, had not from her own defective eyesight set her horizons so close about her; circumstances, childlessness, had imposed them, and, that being so, she had taken up her place in the narrowed circle with resignation so cheerful that it could scarcely be called by that rather depressing name. In fact the gentle old lady was put on a pedestal in the girl's mind, and offerings of incense were made her, a position which now and then she found slightly embarrassing, for Helen in her first moment of understanding, and in the reaction from her previous hasty and mistaken judgment, was one torrent of warm-hearted sympathy, and was disposed to magnify into heroism the performance of those common tasks, just because she had before labelled them trivial.

But from home—she must begin taking up her own little burdens at once—there came no word for her. She herself wrote regularly to her father, but morning after morning passed bringing its posts, and still no answer came to her. Once she saw among the letters laid out for Aunt Susan one addressed in the brisk scholarly handwriting, and could not help glancing at her aunt's face as she read it. But she said nothing to Helen, and replaced the letter in its envelope with a troubled little sigh. Martin also, she knew, had heard from him, but there had been no message for her, no mention even of her. This omission, this intentional disregard of her, though it hurt her, made her sorry also, not for herself, but for him. It was inhuman, but she knew that it was the depth and earnestness of his feeling about her engagement that made him inhuman. On the other hand, she heard constantly from Frank, who hinted that if not a day, at any rate a season might be ever so vaguely indicated to which he could look forward.

The term was drawing to its close, and Martin would go home in a few days' time. It was understood that Helen would go with him, and as the day of departure got near, she knew that her decision must be made, so far as it concerned herself, as to whether she should put off her marriage for some definite time, and do the daughter's part to her father, living at home, obeying him, performing her parish duties as before, making *amende,* as far as she could, for the great act of disobedience which she was going to commit. Practically she did not see the use of it; no good, as far as she could judge, would come of it, yet in a way Aunt Susan was right, the meaning of it, the sentiment of it, was sound. It would not be easy, it would be full of sustained effort, of sustained self-repression; intercourse would be crammed with misunderstanding, the atmosphere would be full of frictional disturbances, but she saw there would be a certain moral gain to set against this. Also, and this too had a very sensible weight with her, there would be gain to her, in the completeness of which her aunt had spoken. Ever since she had consciously woke to her own individuality, her eagerness for her own improvement and enlargement had been of a very vivid sort. And perhaps the most excellent way of all had been here set before her, to compass that not by working for it, but by apparently limiting, maiming, discouraging it. That was a very simple, very elementary suggestion, yet it had never occurred to her in this connection. And it was, well, less crude than the other method.

The evening before her departure she took the opportunity provided by Dr. Arne's going to his chorusmetres after tea, to talk to her aunt again. It had been a chilly day, touched with the autumnal sadness of early-falling leaves, and early-falling dusk, and the window-panes streamed. Though it was still August a fire burned in the grate, and she sat down on the floor by her aunt's chair.

"Father has not written to me once since I came here," she said. "He has written to you and to Martin I know, but there has never been a message to me. I don't say this in any complaint, Aunt Susan, but what is one to do when that happens?"

Lady Susan shut the book she was reading. She had been expecting Helen to mention this, but unwilling to open the subject herself.

"I know he has not, dear," she said, "and I think it very wrong of him. I have told him so. But don't let it hurt you, Helen. If other people, yes, misbehave, there is never anything to be done except to go on 'behaving' oneself. And never let what other people do hurt you. For nothing can really hurt us except what we do ourselves."

"Ah, but in a way I have done it," said the girl. "At least, it is in consequence of what I have done."

"No, your father is wrong, I think," said Lady Susan, with gentle decision. "And now, dear, as you are going away to-morrow I want to ask you something. You go home with Martin, do you not? And then! Have you made up your mind?"

"Yes," she said. "I will not give up Frank, but I will put it all off till next May. Of course if he wishes, he is absolutely free."

"Ah," said Aunt Susan gently. "It is likely he would wish that, I suppose."

Helen laughed.

"Well, no, not very. But till then I shall live at home, if father will let me, and try in every way to please him."

Her voice trembled a little.

"And I hope he will accept that," she said. "And I hope he will be good to me and forgive me."

Lady Susan stroked her hair in silence a moment.

"You have chosen right, dear," she said.

CHAPTER IX.

HELEN was sitting again at the deal table in the "Room," trying to balance the accounts of the quarter. A money-box, cheap but not strong, probably made in Germany, with a florid ornament of tin tacked on round its maw, stood open by her left hand, and on the table was a heap of money consisting chiefly of pennies and small silver coins—the subscription to the "Room" being threepence a quarter—while by her right hand was a pile of equally mean bills, chiefly ending with a halfpenny, for brown holland, cotton, slate-pencils, needles and gum. There was a discrepancy somewhere of ninepence, but add and subtract as she would, that ninepence held its ground like the remnant of the Old Guard. Had it been only deficit, the remedy from her own pocket would have been easy, but unfortunately there was ninepence too much, and though her conscience would not have made any protest at her supplying it, it did not permit her either to pocket it, or to forge a non-existent bill. And all the time her natural impatience, mixed luckily with a certain sense of humour, said to her, "Is it possible to conceive a less profitable way of wasting time than in trying to make ninepence vanish?" Her father, however, with the attention to detail which was so marked a characteristic of his, always looked over the accounts afterwards, and whether there was a discrepancy of a thousand pounds or a penny it made no difference; the principle of admitting discrepancy was equally dangerous in either case.

The Twins had been at home, in a state of total eclipse, for two days of ominous parental silence. Mr. Challoner as usual was busy, Helen was busy also, for after her absence there was more than enough at present to occupy her day. But she had not yet broached the subject that was at the root of the silence: until the skies cleared a little she felt absolutely unable to do so. Her father also had said nothing about it; they ate, they drank, the weather was mentioned and the danger of trouble in the East. Mr. Challoner himself, except when he read prayers, had hardly said half-a-dozen words in Helen's presence: it was "good night" and "good morning" and both were bad. Martin also was, so to speak, in prison, though not like his sister in the condemned cell; he read Demosthenes in his father's study while the latter was writing his sermon, fell asleep and was detected, awoke, and wrote a futile supererogatory set of Greek iambics containing several false quantities, and forms of aorists previously unknown and very interesting.

This morning Helen had received a letter from Frank that troubled her, for he pressed, where he had only hinted before, for some definite sort of date. Reasonably enough, he saw no cause for delay, he knew that in spite of her father's feelings she had accepted his devotion; that was all hers, waiting for her to reward it. The tone was not querulous; if it had been, the letter she must write would have been less difficult; it was simply and sincerely trustful. But before she wrote, she must talk to her father; that could be put off no longer.

For the moment however the "sad mechanic exercise" of the accounts occupied her attention. But though the superficial brain which was employed on addition had its work before it, all that was round her, the walls, the floor, the aspect of the room, the neat new brown-holland covers of the library took that part of her brain that really felt and lived back to the day when she sat there last. The map of geological strata was there, too, with its auriferous belt, and she remembered very well Frank's words about that, "There is a gold-bearing vein in all we are set to do. The trouble is to find it." Yes, indeed, that was the trouble. She did not rebel against the superfluous ninepence, except indeed humorously, but what seemed to her such hard and barren rock was the living in this hopeless silence. Her conscience, her whole sense of moral obligation had accepted the principle indicated to her by the dear aunt—sofa-cushion no longer—of this wider self-completion to be attained by behaving rightly in all relations of life. But at present she had been throwing good money after bad. The dutiful daughter had come home, but no more notice was taken of her than of a mended window-pane.

Mr. Challoner always opened doors smartly. Thus, when the outer door of the "Room," which gave on to a small lobby where wet coats were hung, gave a quick rattle of latch, she knew, with the same certainty as she had known the crisp foot on the gravel, who came.

"Have you finished the accounts?" he asked.

"I can't get them quite right, father," she said. "I think——"

"You have the bills and the receipts, have you not?" he said. "Where are they?"

Helen resented this, but silently; no shadow of it appeared in her face or voice.

"They are all here," she said. "I have ninepence more than I should."

Mr. Challoner sat down and counted up the silver and pence, arranging them in neat shilling heaps, with all the care he would have given to a total of millions. Then rejecting her addition, he added up the receipted bills, and her mistake, one of pure carelessness, was patent.

"That balances then," he said. "Perhaps I had better do the accounts for the future. If I have to do them in the long run, I may as well do them at once, instead of wasting your time over them."

Helen stood up, her resentment showing itself a little.

"Certainly, if you prefer," she said.

He did not answer, but ran a metal clip neatly through the receipted bills, and swept the coins back into the money-box. Then he turned to her quickly.

"What do you intend to do, Helen?" he asked. "As your father, I think I have a right to ask you, since you have shown no sign of wishing to tell me."

The gulf between them seemed to her at that moment immeasurably wide, and his tone was harsh and cruel; it cut her, but cut like a blunt knife with sawing and tearing.

"Father, don't speak to me like that," she said. "I can't bear it, and it does no good. I am trying, and I am going to continue trying, to do my duty to you."

For one moment the sternness vanished from his face.

"You are going to give him up?" he asked.

"No, but I am going to live quietly here, if you will

have me, for the next six months," the said, "doing my work in the parish just as usual. During that time I will not see Frank. If you wish, I will not even write to him, except just once."

She sat down again opposite him.

"I want to do something for you, which is hard for me," she said. "I want to make you believe that I am trying to be a good daughter to you. I know we disagree vitally and essentially. But is that any reason why the dearness of our human relation should be diminished?"

Her voice sank, but looking at his face she could see that the momentary brightness as he asked the last question had vanished again, and he sat looking not at her but out of the window, without replying.

"Father," she said gently, "I have spoken to you."

He shook his head, then looked at her.

"It is useless," he said.

Then suddenly the chilling reserve and silence of the last days gave way like ice before the south wind.

"My God!" he said, speaking more to himself than her, "What have I done? What have I done? Has this come for some dreadful fault of mine of which I am ignorant? All your life, Helen, I have tried to train and teach you in the knowledge and fear of God. As He sees me, I have done my best, according to my lights. Never once to my knowledge have I not prayed every day that His blessing should guide and illuminate every step you take. And I cannot believe—that is my difficulty—that you try to follow His will in this. It is impossible that——"

He broke off suddenly, with a sudden helpless raising of his hands, indescribably pathetic.

"God help us both!" he said.

There was a long silence, and his fingers clenched and

unclenched themselves as he sat staring dismally out of the window. All her life, as he had said with absolute honesty, he had tried to bring Helen up in the knowledge and fear of God; and this decision of hers, from which he now realised he was powerless to move her, was like some overwhelming blow struck at him from the dark. He could not understand, he could not even conjecture in the vaguest way, what it meant, or how he was meant to take it. In sorrow, renunciation, bereavement, it was at any rate possible to acquiesce in there being a design. But that his child should do this was inexplicable; it could not be the will of God. Something of this Helen read in his face; and she saw, for the first time fully, how the blow had staggered him. His strength had given way under it; all vehemence and anger was dead, and dead too was the hope that she would come round to him. He was helpless. And the strangeness of that in one so certain, so accustomed to go without hindrance or obstacle along the straight road of his God-fearing life, touched her with a profound pity; so that for a moment, had he but known it, her decision flickered and wavered like a candle-flame blown about in a draught. She questioned herself whether such suffering could be right, whether that which caused it could be justifiable, whether at whatever cost to herself or another she could permit it to be. It was like the suffering of some animal, blind, uncomprehending, a thing intolerable. And the animal that suffered was a strong man and a wise and her father.

She sat down on the edge of the table beside him.

"O poor father! poor father!" she said.

He looked at her with a wretched semblance of a smile.

"Ah, that is not the point, Helen," he said. "What I

feel, all my pain, is nothing, nothing. Why I feel it is everything, dear. O you poor girl, blind, blind!"

Then at last that tie between father and daughter or mother and son, one of the immutable and indestructible things of the world, stirred, vibrated, made music, and for a moment across the infinite gulf between them their spirits and their hands met.

"Dear girl," he said, "it will be delightful to have you at home. I was afraid that those happy days of work, you and I, side by side in this home, were over. I thank you for that, Helen; your father blesses you for that. Stop with me as long as you can. How long you—and he—must settle. And, my dear, I am so selfish as to take your offer fully. Do not see him or write to him. Perhaps——"

He paused a moment, stroking her hand.

"And try to make allowance for me," he went on, "when I am hard or gloomy, or out of spirits. But I am so utterly at sea: my landmarks have gone. I don't understand. I can only pray that you and I may have light. God bless you, my dear, now and always!"

Helen wrote the same day to Frank.

"My Dearest,

"I have just come home, and I have settled to do a thing which is very hard on both of us, but I cannot do otherwise. Frank, we cannot be married yet. We must put it off for six months, or seven is it not?—till next May. And for six months I must live quietly at home here, and not see you. There, it is written. This too: you are absolutely free—ah! in spite of all these troubles I can't help smiling when I write that.

"But I can't act otherwise. My father is in a state of

misery about it which I can't describe to you. Somebody he loves is deliberately—this is how he sees it—going to do a wicked thing. This morning when he talked to me about it, I wondered whether I could be right in continuing our engagement at all. But I can't give you up. My love for you is the best part of me, and the most living part. You see I *am* yours. Oh, my dear, if only things had been otherwise, if you could believe! If you could only have not told me, have let me think you were a Christian. No, I don't wish that really. It would not have been you.

"He is my father. All my life he has watched over me, prayed for me, loved me. Even if he had been a bad father I should still have owed him all I am, until the day I met you. And the only way in which I can repay him anything is by doing this. It is small change, I know, for all his gold, but it is all I have. At least then and at most I must do it. I must stop here with him—he was such an old darling when I told him—trying to be cheerful, trying in little tiny human ways to be a good daughter to him. And it is all so infinitesimal: it is as if I gave him remedies for a cold in the head when he had cancer. I feel so mean in offering him so little. But there is only one other thing that I could offer him, and that I cannot. And, indeed, though this looks so little, and makes little show, it costs me something. It does indeed.

"And I must do something more. I think I must not even write to you. While I am here I must have no connection with you. It would be incomplete without that. One letter you must send me, when you have thought this over, to say that you agree with me if you can.

"And if you cannot? I must do it all the same.

"Do you remember telling me of Magda's cry? That,

too, tells me to do it. I should be stunted, selfish, if I did not.

"Ah, Frank, my darling, be good to me. I long for you every day, and it is going to be so awfully dreary without you. "HELEN.

"I walked through the wood to-day where you set the hare free. I shall walk there every day. And I looked at the geological map with the 'auriferous reef' in it. Martin is here."

The letter was not difficult to write, though the final determination to write it was so hard that when it came to the paper and ink she sat long with pen undipped, unable to begin. But the memory of the bewildered misery in her father's face that morning as he sat looking out of the window in the Room had given her a real sense of responsibility towards him. It was her business to find some anodyne for that; perhaps, the proof before his eyes, kept there day after day and week after week, that she wanted to do her best—might serve. Anyhow, at the moment it had awakened his humanity and his fatherhood: his hand had reached to her across the gulf; two puzzled blind folk had clasped hands in the darkness.

Nor was the waiting for Frank's answer difficult, she knew him so well. And she was not disappointed here—the very brevity of the reply was honey to her.

"DEAREST,

"You must do as you must do. Magda says so, and so do I. But I am rather low, though she tells me not to be."

"FRANK."

But it was then, when she had made the difficult determination, and Frank had so ungrudgingly consented, that Helen's difficulties began. Each day was an endless series of infinitesimal knots, not to be cut, but each to be patiently, cheerfully unravelled. Each singly she could tackle, but she had to avert her eyes from the future, for the series of knots stretched into dim distance. All day too there was with her the desire to see Frank, just once to see him, and perhaps cry a little on his shoulder: all day too there was the face of her father, always sunless, always grave. He had never, it is true, been other than austere in his domestic life, but then Helen had always known how deep was his love for her. But now it seemed to her sometimes as if he was trying to stifle and extinguish it; that, knowing as he did, that there was soon to be an irrevocable rupture between them, a rupture that would divide them further than death divides, he was schooling himself to get used to it, as a man may school himself when he sees one he loves in the pangs of mortal illness to adjust himself beforehand to the loss that is coming. The marks of his suffering too were pathetically plain, and again and again she asked herself whether she had not only increased it by doing that which cost herself so much. Was it only an impulse of barren sentimentality that she had followed? Was she like a surgeon who gives an ineffective anæsthetic which should not deaden or mitigate the wrench and shock that was coming?

The encouragement she could find was but small. But it was this, that in any case she had done what was most difficult, and what seemed, not only to her but to Aunt Susan, to be right, and as such was fully accepted by her lover. Yet what if after all this was a mere senseless mutilation of herself, an objectless asceticism?

It was this doubt that day after day most troubled her. Had she seen the least sign of bud on the barren stem she would have been much more than content. But the days became weeks; and there was still none, not even any return of the moment's tenderness her father had shown at their first talk. She could not see that any practical good was coming of her renunciation. Like a wrecked sailor on a raft, she watched, as for a sail, for any horizon-distant sign that her father accepted her marriage and gave her credit—though she did not want the credit herself, but only longed for the evidence of it—for doing her best. But there was no such sign. He continued to use the prayer for Turks, infidels and heretics.

What made things worse was that Martin, the beloved twin, with whom disagreement was a thing unthinkable, radically disapproved of what she was doing; and his disapproval, she was afraid, was terribly practical, namely, that it was quite certainly no use. Two things, however, after some three weeks of what seemed fruitless endeavour, kept her to it. One was a letter from Aunt Susan, to whom she had sent a despairing sheet, containing a memorable sentence: "God does not always pay on Saturday, Helen," she had said. The other was an innate pride, that forbade her to accept defeat. Here she feared also to lose the respect not only of her father but of Frank.

"Yes, my darling, you tried it," she imagined him saying, "and you found it was doing no good."

And that he should say that was somehow intolerable to her. Whatever she might be, she would not be feeble. "The lame and the blind that are hated of David's soul" seemed to her a very legitimate object of detestation.

She would not give a thing up because she mistrusted her power of doing it.

Thus her apparent failure consumed itself. With the divine confidence of youth, the less successful she seemed to be the more she spurred herself on to strive. All her sense of right had told her, when she made her decision, that she would thus be doing her best; her judgment was arrived at coolly and sanely, and the present practical ill-success of it argued nothing against the principle.

Then came a crowning despondency and agitation in something Martin told her after he returned from a visit to Lady Sunningdale. The short history of that visit, however, claims an episodic precedence.

Lady Sunningdale had sent her motor over from Fareham to fetch Martin, and when he arrived about tea-time he rushed straight out on to the lawn to find her, but only encountered the chilling looks of several total strangers who were talking about fiscal problems, and seemed surprised, if not pained, to see him. This was discouraging, and he was wondering what place there was to flee unto, when a footman came out after him to say that her ladyship was in her bedroom and wished to see him there immediately. Martin could not help giving a little giggle of amusement at this; and the footman, preceding him upstairs, threw open the door and announced him.

The room was large and very rose-coloured, on the principle of Lady Sunningdale's famous maxim that bedrooms should be optimistic. She herself was reclining on the optimistic silk coverlet of her bed, with her shoes off and the blinds down.

"Is that you, monster?" she asked. "I am an absolute wreck. Yes, pull up one blind and sit down at a respectful distance. Martin, you must promise to play absolutely all the time you are here, like a barrel organ, or I shall die. I shall send a footman to you after each tune with twopence on a tray, and orders that you are not to move on. The house is crammed with perfectly dreadful people; I cannot imagine why I asked them. I hope you have not brought your gun, because I shan't let you go shooting. You will have to talk to me all day, except when you are playing. Don't tread on Suez Canal or you'll be drowned. Frank is here and Stella. Otherwise—my dear, why are politicians so impossible? And why is Helen behaving like a mad-woman? Really I thought she had more brains."

Martin had pulled up one blind during this, and revealed the room. There were pink-silk walls, on which were several pictures of Lady Sunningdale of not very recent date, a pink carpet, white furniture, and a particularly large and pink bed. Lady Sunningdale, fenced, like Egypt, on the one side by Suez Canal, and on the other by Sahara, was lying propped up by a quantity of huge pillows and cushions. French books with yellow covers bestrewed the bed, and fragments of chewed pages suggested that the dogs had eaten one, like Jezebel, leaving only a few very indigestible pieces. A French maid hovered uneasily about a toilet table, and appeared to be putting things in drawers. Considered as a wreck, finally, Lady Sunningdale looked particularly large and seaworthy.

"Miss Plympton is here?" asked Martin in an extremely disengaged voice, but with his face suddenly infected by the prevailing optimism.

Lady Sunningdale drew conclusions before most people could have arrived at data.

"Yes; ever since you played to us at Chartries she has been trying to learn the 'Merry Peasant,'" she said. "She is not getting on very well, but art is long, is it not? So is life. Too long, I think, sometimes. But, my dear, the rest of them! They talk about fiscal problems and what they've shot. Even Frank appears to be vaguely interested in free trade or free food or free drinks or something, which is deplorable of him. I expect him here immediately. My bedroom is the only place where one can be free from those intolerable bores. There are three, three Cabinet Ministers in the house! Really, politics ought to be considered a dangerous habit, like morphia. In fact, there is a very great resemblance between them. They are both drugs that send one to sleep, and the habit grows on one: you have to take more and more, and the result is death of the intellect, which is quite as lamentable as death of the body, and renders you far more tiresome to other people: for after all, when one's body is dead, one is put away. But people whose intellect is dead are not put away at all: they pervade society. There is no one in the world so lost as the intellectually lost. How big Hell must be! Talking of that, how is your father? What a Bear!"

Martin had settled himself in a rose-coloured chair, and gave a great shout of laughter, suddenly checked.

"Quite well," he said. "He always is."

"Yes, that is so like him," said she. "But really have you any strain of insanity in your very extraordinary family? My darlings, did I kick you? O Sahara, naughty! All that book, and I hadn't read it. *'Commandez du thé, Hortense.'* So convenient, she doesn't know a word of English. Did you ever see such a murderish-looking woman? But she can make hats out of a toothbrush and some waste paper. Some day she will kill me for my diamonds, and find out afterwards that they are paste. Then she will be sorry, and so shall I. Do attend, monster. Can you tell me why Helen, head over ears in love with him—that was why I brought them together—should behave like that? Shutting herself up with the Bear and that dreadful aunt of yours who plays Patience. And Frank thinks in some confused way that it is so beautiful. He looks so funny when Helen's name is mentioned, rather like a widower who hears a hymn-tune in four sharps on Sunday evening. So frightfully old-fashioned that sort of thing. Those two find a sort of spiritual thrill in standing a hundred miles apart and shouting 'Caro mio! O carissima!' to each other at the tops of their voices. I can't bear that sort of Platonic love. Yes, you Challoners are all mad. If Becky Sharp lived with Savonarola in a grand piano, you would find a little Challoner crying on the drawing-room carpet one morning."

"Why Becky Sharp?" he inquired parenthetically.

"Only to add a little *joie-de-vivre*. No imputation on your morals."

Lady Sunningdale struggled to a sitting attitude on the bed. Several French books flopped to the ground, and were instantly worried by the dogs; *Zöhár* and *A Rebours* flew in gnawed fragments about the room.

Martin agreed with Lady Sunningdale in the view she took of Helen's conduct, but he felt bound to defend his sister against so wild an attack.

"Anyhow, she's doing a difficult thing because she thinks it right," he said. "Give her credit for the difficulty."

"Difficult?" cried Lady Sunningdale. "There is no

merit in doing a difficult thing just because it's difficult. I might just as well try to stand on my head in the drawing-room, and say to my wondering guests, 'Admire me, please. Though foolish, this is difficult, and is only accomplished by prayer and fasting.' Is that profane? I think it must be, because my father was a Nonconformist, and whenever I say anything without thinking, it is nearly sure to be a reminiscence of my unhappy childish days, and comes out of the Bible. But it doesn't prove that a thing is the least worth doing because it is difficult. She is standing on her head then? And in a Parsonage too!"

"Yes, it amounts to that," said Martin. "But with a moral purpose."

There was a discreet tap at the door, and Hortense entered with tea.

"Ah, muffins," said Lady Sunningdale in a mollified tone. "The under piece, please, Martin. How delicious! But though I am not cynical I always a little distrust moral purposes. If you do a thing with a moral purpose it usually means that you do it, because, if you didn't you would be uncomfortable inside. Good people are such cowards, they are afraid of a little pain in their consciences. To avoid that they go and act in some foolish, antiquated manner, and everyone says, 'What a saint!'"

Then out of all this nebulousness, like the gathering clouds of a thunderstorm, there leaped a sudden flash, like lightning and rather like genius.

"She is doing sacrifice to an ideal she doesn't fully believe in," she said. "Helen doesn't believe in certain things as your father does. Else she would never marry Frank at all. She would have screamed loudly for help when he asked her, instead of saying Yes. Her sacrifice, therefore, isn't quite sincere."

Then a sort of confusing roar of thunder followed, marring the sharp conclusiveness of the lightning.

"I cannot bear seeing people making a mess of their lives," she said, "and it is such a pleasure to see them make a really clean job of them. Yes. Why continue poking round in a Parsonage when you have made up your mind to go away? It is like ordering the carriage to go to the station, and then for no reason saying that you will go by the next train. She has shattered the happy Parsonage life, and is feebly trying to pick up the bits, instead of ringing the bell for somebody else to pick them up and leaving the room. It is silly."

"Ah, Helen is not silly," said her brother.

"I did not say that. Yes, slap Sahara twice, hard. But I said she is doing a silly thing. Now I am silly, but I hardly ever do a silly thing. Yes, come in. It must be Frank. Sunningdale never knocks, and nobody else ever comes in."

Frank appeared at the door.

"I was sent for," he said apologetically. "Ah, Martin!"

That rang true. "You are her brother" was behind it, and the romantic touch did not escape, though it rather irritated Lady Sunningdale. Personally she disliked romance, on the general grounds that in real life it was old-fashioned. To her the two completely satisfactory methods of expression were melodrama and farce And Frank's greeting to Martin, the hand on the shoulder the linked arm, was all romantic, and just a little tiresome.

"Frank, what have you been doing with yourself all

day?" she cried. "I have not set eyes on you. But of course if you do prefer gold and Chinese labour to my inspiring conversation—yes, help yourself to some tea. And all the muffin there is."

But Frank still lingered by Martin.

"How is she?" he said. "Is all well? Any message for me? No, of course there can't be. She meant that. But she is well?"

He sat down on the foot of the rose-coloured bed.

"Dear lady," he said, "I have done both. I went out playing golf with a Colonial Secretary, I think, and we talked about fiscal problems. Then I drove off into the bushes and lost the ball. So I said, 'Will the price of golf-balls go up?' Then he drove into the bushes too, and he said, 'I expect so. So we will not look for them for a year. They will then be more valuable than they are now, but will require painting.' Lucky golf balls! The longer most of us live the less valuable we become."

Lady Sunningdale rather resented this.

"The older people become the more paint they want," she said, "but the other is absolutely untrue. Until people are of a certain age they are of no value at all. I hate boys and girls. You only just escape, Martin, and I don't think you would unless you could play like an elderly person. Young people want airing; they want to be out in the world for a time to get ripe. Tact now—tact and good temper are quite the only gifts worth having, and tact is entirely an acquired quality. Until all your edges are rubbed down, you cannot have tact. People with edges are always putting their elbows into others, instead of rolling along comfortably. You have no tact, Martin, and Helen it appears has less."

Frank held up an appealing hand.

"Ah, please, Lady Sunningdale," he said.

"Dear Frank, it is no use saying 'please,'" cried she; "Helen is behaving idiotically. She ought to have smoothed the Bear down somehow, and deceived him for the sake of his comfort. Martin I think would deceive his friends to make them comfortable. Considering how dreadfully uncomfortable life is, the first duty towards our neighbour is to try to make things pleasant. You too, Frank, you have no tact. You ought to have said the Ten Commandments, or whatever it is, very loud in the vulgar tongue when you went to the Bear's church, and then there wouldn't have been any question at all. I would be a Parsee or a Plymouth sister to-morrow, if it would make Sunningdale groan less. He has taken to groaning; I suppose his mind hurts him, as he says he's quite well."

"Did you say that I would deceive people to make them comfortable?" asked Martin.

"Yes: at least I hope you would. But you Challoners are all slightly cracked, I think. You owe your vividness to that. You, Helen, your father, all see things out of their real proportion."

"Have you ever seen Aunt Susan?" asked Martin.

"No: is she dreadful?"

"Not at all, but not vivid. It was she who really made Helen go home and live there."

"Then your Aunt Susan is a very stupid person," said Lady Sunningdale. "My dear, there are only two sorts of people in the world, the clever and the stupid. Nobody is good, nobody is bad. At least they may be for all that it matters, but goodness and badness in themselves have no result. There is nothing more colourless than moral qualities; it is only brains that give colour to them. Do you choose your friends because they are good? I am sorry for you. Of course I don't want you to choose them because they are bad: the one is as idiotic as the other. But brains! There is nothing else in the world, and very little of that. And moral qualities are like corsets. If they are tight they hinder free development, and if they are loose you might as well not wear them at all."

Lady Sunningdale had taken her feet off the bed during this remarkable speech, and looked more closely at Martin.

"Your forehead is bulging, Martin," she said, "and your hair is dipping like a plume into your left eye. That happens, I notice, when you play, and it means you are thinking. So you are thinking now. What is it?"

Martin did not deny the soft impeachment.

"Yes, I was thinking," he said. "I don't imagine that what I was thinking about would interest you in the least."

Lady Sunningdale made a gesture of despair.

"Haven't you grasped the elementary fact," she said, "that anything anybody thinks about is deeply interesting? All the events of the world—who said it?—take place in the brain. Sahara, darling, I am not a mutton-bone, nor are my rings good to eat. Suez, how tiresome! And I hadn't read a page of it! Yes: what were you thinking about, Martin?"

Martin lit a cigarette from a smoked-down stump before he replied.

"I was thinking whether I was going to join the Roman Church," he said.

Lady Sunningdale gave a deep, contented sigh.

"That's the sort of thing I really like," she remarked. "Poor Bear! Now why, why, why do you want to do that? Yes, turn Sahara out, Frank, she is so restless: Suez Canal always follows her. And shut the door. Now close your eyes and think, Martin, for a minute, if you like, and then tell me why."

Frank said under his breath, "I thought so," and returned to his chair almost on tiptoe. Martin did not close his eyes at all, but looked at him.

"Frank knows why, I expect," he said, "though I haven't hinted it to him till this moment. Why is it, Frank?"

"Well, in one word 'Beauty,' said he.

Lady Sunningdale was completely bewildered.

"Incense? The Virgin Mary?" she suggested vaguely.

Martin frowned: for a moment he looked exactly like his father.

"Ah, what is the use of my telling you, if you say that sort of thing?" he asked.

"But I really haven't an idea," said she. "Did I say anything dreadful?"

"Frank, speak: you know," said he. "I never know what I am talking about when I begin to talk."

"It is only a guess."

"You have guessed right. I believe you are always right."

"Well, get on, somebody," said Lady Sunningdale with a show of impatience.

"All is beauty," said Frank, "and knowing this is love, and love is duty."

He smiled across to Martin.

"You quoted that, you know, to Helen," he said, "on the day your father found 'The Mill on the Floss.'"

"What did he find the Mill on?" asked Lady Sunningdale. "Oh I see. George, Eliot, isn't it? How dull! I read a book of hers once, 'Scenes from Something,' and thought it so like your father's house, Martin. But all is beauty, is it? I should have said almost everything was ugly. Anyhow what has it all got to do with the Pope?"

Lady Sunningdale's discursiveness, the reader will have noticed, was liable to put in an appearance at any time, even when she was really interested. She herself explained this by the fact that she never thought about less than three things at once. Consequently, when she opened her mouth any of the three was liable to make its escape.

"Yes, that is it," said Martin, answering Frank's last remark. "I am a Christian, and I cannot any longer be of a Church that leaves out beauty from its worship. Why, if you love a thing, if you believe in a thing, you must approach it through beauty, it seems to me."

He paused a moment, and then the words came as they had never come before. A sudden clearness of vision was his; he saw his own thought with precision, and he could at that moment of self-revelation delineate very accurately.

"Why, when one's friends come to see one," he said, "one makes the room tidy; if you came to see me at Cambridge, Lady Sunningdale, I should take down my pipe rack and put it in my bedroom, I should sweep my hearth, I should give you a clean tablecloth for lunch, I should get flowers for the table, I should practise something which I thought you would like to hear me play. I should in my small way put all the beauty at my disposal at yours, and put the ugliness away. But—but take Chartries church. How beastly!"

Martin paused a moment: Frank was observing him quietly from underneath his hand, for the afternoon sun was pouring its light from the window where Martin had pulled up the blind full into his eyes. The boy seemed to him at this moment suddenly to have grown up, become vivider, to have thought for himself. Crude, elementary, unconvincing it all might be, but it was original. And Martin's next words endorsed his opinion. Certainly he was not a child any longer.

"How dare they? How dare they?" he cried. "A wheezy organ: awful woodwork: terrible windows! Is there anything more hideous in all England than Chartries church, unless it be a county jail for the confinement of prisoners? Because it is for God, will anything do?"

There certainly was crudity here: Frank felt that, though Lady Sunningdale did not, for her indifference on religious matters was perhaps the profoundest thing about her. He had inquired and rejected, she had never even looked in that direction. Martin had inquired too and found an awful Presence. And he was ashamed to call in old clothes, so to speak. What was at the service of God was his best. All that was not best was an insult. And his face flushed suddenly.

"Why, if that church was my room, and you came to see me, I would cover up the stained glass," he said. "I would make it decent. I would, I would——"

He paused for a moment, then found the word.

"I would have 'form,'" he said. "I would give you politeness. I would not say, 'She knows me; she will understand,' and sit with you in a back-bedroom, slops about, tooth-brushes, anything. But because God understands, are we to say, 'Anything will do?' Why, when the Queen came to Chartries, we had four courses for lunch, and a red carpet."

He broke off suddenly.

"Do you understand what I mean?" he demanded of Frank.

Frank understood perfectly, for he had known a long time, what Martin had only just learned, that "form" governed his life. For he did and always had done everything he believed in as well as he could do it, lavishing thereon all the pains and trouble at his command with the instinctive open-handed generosity of love. These pains he did not bestow grudgingly, nor count the expenditure; whatever was worth doing was more than worth all the pains he could possibly bestow on it. That impulse lies at the root of every artistic temperament, endless trouble for ever so minute a perfection, ever so infinitesimal a finish. But Frank, like an equitable judge, had to state the other side of the case to Martin.

"What will your father say to it?" he asked, using the most commonplace phrase.

Martin looked at him quickly.

"Same as he said about you and Helen," he remarked.

Lady Sunningdale could not help a little spurt of laughter, the *repartée* was so exquisitely simple. But she checked it at once.

"But it's too awful for him," she said. "First Helen and then you. Martin, do you think you ought——"

"I don't know, but I must," said Martin.

"But it doesn't hurt you to play a creaky organ. And the stained-glass windows don't hurt you."

Frank had seen further than this.

"How necessary do you feel it?" he asked. "That is the whole point. Is it as necessary as—as Chopin?"

The door opened, and Hortense entered.

"Sept heures et demi, madame," she said.

Lady Sunningdale started to her feet.

"Monsters, you must go at once!" she cried. "Yes, dear Martin, it is *too* interesting! You will play to us this evening, won't you? So glad you could come, and did you ever see such a mess as the dogs have made? But those things don't hurt you any more than brushing one's teeth hurts, though it cannot help being a terribly inartistic performance. And you ought to consider Helen as well. Not that it matters what Church one belongs to as far as I can see. Sunningdale might become a Parsee to-morrow if it would make him any happier, only there really is no sun in England, so I don't see what he would worship. How nice always to sit in the sun and say one was worshipping! Yes. You extraordinary boy, fancy your being religious in your little inside. I should never have guessed it. But you got quite pink when you talked about Chartries church. Most religious people are so dull. Is that a dreadful thing to say, too? Dinner at eight. Take him and show him his room, Frank."

Lady Sunningdale certainly had the knack of bringing quite unique combinations of people together, and of making them behave quite characteristically of their respective selves. She herself—this may partly account for it—behaved with such childlike naturalness that it was quite impossible for those with her to be self-conscious. As a hostess she was quite incomparable, for, rejecting all known conventions which are supposed to be binding on that very responsible class, instead of behaving to each of her guests as if he were a mere unit in the colourless mass known as society, she talked direct and unmitigated "shop" appropriate to each. To-night there was present among her guests a traveller in Central Thibet, to whom she talked cannibal-shop, so much encouraging him that his account of his adventures became scarcely narratable; an astronomer who knew Mars better, it appeared, than the majority of dwellers on this terrestrial globe know the county in which they live; several Cabinet Ministers who received relays of telegrams during dinner (always a charming incident), their wives, whose main preoccupations were appendicitis, golf, and babies; a duchess of American extraction who shied violently when the words "pig" or "Chicago" were mentioned; and a German princess who, when directly questioned, seemed doubtful as to where her husband's principality lay, and was corrected on the subject by the astronomer. But owing, perhaps, to the advent of the Twin (the name by which Lady Sunningdale referred to Martin), though she had previously confessed that she found her guests "dreadful," to-night she went bravely ahead, steering a triumphant course over shoals where she grounded heavily, and dashing onto rocks that should have made a wreck of her. The dinner table was round; she herself set an excellent example by screaming over smilax and chrysanthemums to the person most distantly removed; and Babel, that god so ardently worshipped by hostesses, shed his full effulgence over the diners. Thibet and the Chaldæans easily led on to astronomy, astronomy to the observatory at Chicago, which occasioned a sudden and thrilling silence, and from the United States it was but a step to fiscal problems, in which all but the Cabinet Ministers laid down incontrovertible opinions. Then golf let them into the circle again, and the story of a golfer being carried off the first tee after a futile drive, and expiring an hour later from an operation for appendicitis, while his wife was being confined, was charmingly to the point. In fact, the desultory rapidity of conversation left nothing to be desired, and all was due to Lady Sunningdale's inimitable plan of talking shop to the shopkeepers.

Later, Martin played; there was Bridge, and Lord Sunningdale as usual went to sleep, and on awaking revoked, subsequently explaining the revoke to the satisfaction of everybody but his partner, who remained dissatisfied to the last. Women took bed-candles, men gravitated to the smoking-room, though, since everyone had previously smoked in the drawing-room, this seemed unnecessary. But the fact is one without exception; men left alone leave drawing-rooms alone.

Soon again, after the long day's shoot, the smoking-room yawned itself to bed, and Cabinet Ministers, the traveller and the astronomer being gone, Frank was left alone with Martin. There was no design in the matter; both hated going to bed as much as both detested getting up, and they were neither of them sorry to have the opportunity of more talk. Frank had got up from his chair on the last exit, took a whiskey and soda, and moved to the fire-place.

"Lady Sunningdale is extraordinarily clever," he remarked, "but I can no more discuss anything with her than I could with a dragon-fly. She is always darting."

Martin laughed.

"Go on, then," he said.

Frank sat down.

"Are you determined, Martin?" he said.

"I think so. I don't see what else I can do."

"I asked you a question before dinner, which you didn't have time to answer. Is it as much to you as Chopin?"

"Why do you repeat that?" asked Martin. "It does not seem to me apt. How can I make such a comparison?"

"Easily, I should have thought."

Again Martin's likeness to his father started to his face.

"You say easily," he said. "Take this then. What would you do if, in order to get Helen, you had to tell a real mortal mean lie, the sort of lie that would make you blush in the dark?"

"It's like that, is it?"

"Yes, just like that. I must. I can't tell you why. I don't know whether I know, except as regards what I said in Lady Sunningdale's room, that if in anything, in worship above all is beauty necessary. That is true, but it is only a sort of symbol of what I feel. Other people feel differently: they are less materialistic than I, and ugliness doesn't get in their way. But if you happen to be gifted or cursed with the artistic temperament—Lord, how priggish that sounds!—I don't see how you can help demanding beauty in the service of what is sublime."

"I never knew you thought about these things," said Frank, rather lamely.

Martin snapped his fingers rather impatiently.

"More fool you, old chap," said he. "All the same, I don't see why you should have. So I'll apologise. Probably you thought that because one has high spirits, a really fine capacity for playing the fool, and also a certain leaning towards the piano, that I never took anything seriously. Nor did I till lately. In any case this is really so much more my concern than anybody's. I've got to lead my own life, not to be dragged about like a sheep. And I must."

He paused a moment.

"I have only given you an external instance of what seems to me an underlying principle," he said. "The difference in 'form' between the two Churches is an illustration of the desire of the Roman Church to enlist beauty in the service of God. That desire is the spirit of Romanism. Now English people, take them all round, are extremely deficient in the sense of beauty, and utterly blind to its importance. And in church I think it really seems to them slightly inappropriate. The Roman Church is mystical, romantic, poetical. The English is Puritan, and ugly, and literal. And do you know, as soon as I began to think, I found I could not stand Puritanism. Heavens, how I have jawed!"

Martin got up briskly from his chair, with the unmistakable air of closing that particular topic. In his youthful, boyish manner there lurked a great deal of masterfulness, which those who came in contact with it might be disposed to call obstinacy. Though he never adopted any attitude so ungraceful as that of a donkey with its legs planted outwards towards the four quarters of the compass, the effect on such as pulled was about the same. If he chose he would smilingly refuse to go in any direction whatever, certainly until all efforts to move him were relaxed. But as he knew himself, and as Frank suspected, there was just one person in the world with whom, hitherto, he had never adopted this attitude, and that was his father. Never yet in his life had he set his will calmly in opposition to Mr. Challoner's. As he had once told his father, he was frightened at him, he feared his anger, but there was certainly no one else in the world whom he would radically disagree with, and yet obey. And some cold intimate knowledge of this had suddenly struck him when at this moment he stopped the conversation. All that he had said he had honestly felt; but vivid as was his imagination, when he flashed a light into his father's study at home, he could not picture himself there saying this to him. His own figure wavered, as if blown by a draught.

There are certain plants which apparently lie dormant, as far as outward observation can go, for months and even years together, and then suddenly grow with an incredible swiftness, putting forth leaf, bud and blossom with a rapidity that is almost uncanny. Some invisible storage of force must certainly have been taking place during the prolonged dormancy, the root-fibre has prospered and been accumulating vitality out of the ken of human eye, transforming the fertile elements into itself, and the visible result is the constellation of sudden blossom. And a similar phenomenon is observable in that most obscure of all growths, that of the human character. There are no clear causes to be registered of this sudden activity; only the essence of the conditions favourable to growth must have been stored within it, till its reservoir has been filled to overflowing, and discharges all at once its potential energy. It struck Frank this evening that some such inexplicable sprouting had just begun in Martin. He had quite suddenly taken a distinct and defined line of his own, and was under the spell of an irresistible original impulse. He had never been, it is true, devoid of vividness or vitality, but he had never yet taken a step. He had been held by the scruff of his neck with his nose to the grindstone of classical education without attempting to raise it, and his recent emancipation had been entirely contrived by others, while he himself had stirred not a finger in it, leaving Frank, his uncle and Lady Sunningdale to fight his battle for him, merely sitting in his tent and, it is true, receiving the news of victory with engaging delight. But now his character showed growth: he had thought for himself, come to a conclusion consistent with himself, and was apparently prepared to act on it.

And now that the growth had begun it was not so hard to see the causes which made it inevitable. For he was an artist through and through; in all his tastes, in all his achievements, the note of "form" sounded trumpet-like. And if, which Frank had not known, the desire and the need of God was in the woof of his nature, that too must be expressed with the æsthetic beauty in which, necessarily to him, emotion had to be clothed. He could be and was slovenly in execution where his artistic sympathies were not aroused, as his more than mediocre performances in classical languages could testify; but where his feelings were concerned, any expression of them had to be made with all the excellence obtainable. He was not able himself to do badly what appealed to him, neither could he watch or take part in a thing that was badly done. And the growth that he had made consisted in the fact that he recognised this.

CHAPTER X.

KARL RUSOFF got up rather wearily from the piano, where he had been practising for the last three hours, stretched himself, and for a few seconds held his fingers against his eyes, as if to rest them. The afternoon was a little chilly, and he walked over to the fireplace, where he stood warming himself. The cheerful flickering blaze shining through his thin long hands, made the fingers look transparent, as if they were luminous and lit with a red light from within.

From the windows the dun-coloured gloom of a cloudy Spring afternoon in London left the room vague and full of shadows that huddled into the corners, while the light of gas-lamps already lit in the street outside cast patches of yellow illumination high on the walls and on the mouldings of the ceiling. The room itself was large, lofty and well-proportioned, and furnished with a certain costly simplicity. A few Persian rugs lay on the parqueted floor, a French writing-table stood in the window, a tall book-case glimmering with the gilt and morocco of fine bindings occupied nearly half of the wall in which the fire-place was set, two or three large chairs formed a group with a sofa in the corner, and the Steinway grand occupied more than the area taken up by all the rest of the furniture. There perhaps simplicity gained its highest triumph; the case was of rosewood designed by Morris, and the formal perfection of its lines was a thing only to be perceived by an artist. On the walls, finally, hung two or three prints, and on the mantelpiece were a couple of reproductions of Greek bronzes found at Herculaneum.

It was a room, in fact that spoke very distinctly of an individual and flawless taste. Wherever the eye fell it lighted on something which, in its kind, was perfect: on the other hand there was nothing the least startling or arresting, and above all nothing fidgety. It was a room pre-eminently restful, where a tired mind might fall into reverie, or an active mind pursue its activities without challenge or annoyance from visible objects. Pre-eminently also it was a room instinct with form: nothing there should have been otherwise.

Karl stood in front of the fireplace for some minutes, opening and shutting his hands, which were a little cramped, a little tired, with the long practice they had just finished. His mind, too, was a little tired with the monotony of his work, for his three hours at the piano had been no glorious excursion into the sunlit lands of melody, but the repetition of about twelve bars all told from a couple of passages out of the "Waldstein Sonata" which he was to play next week at the last of his four concerts in St. James's Hall. And though perhaps not half-a-dozen people in that crowded hall would be able to tell the difference between the execution of those dozen bars as he played them yesterday and as he could play them now, he would not have been the pianist he was if it had been possible for him not to attempt to make them perfect, whether that took a week or a month. The need of perfection, which never says, "That will do" until the achievement cannot be bettered, was a ruling instinct to him.

Besides, to him just now the presence of one out of

those possible six auditors who might be able to tell the difference, was more to him than all the rest of the ringing hall. Sometimes he almost wished he had never seen Martin, never at any rate consented to give him lessons, for in some strange way this pupil was becoming his master, and Rusoff was conscious that the lad's personality, never so vivid as when he was at his music, was beginning to cast a sort of spell over his own. Brilliant, incisive, full of fire as his own style was, he was conscious when Martin played certain things that his own rendering, far more correct, far more finished though it might be, was elderly, even frigid, compared to the other. The glorious quality of his exuberant youth, a thing which in most artists is beginning to pale a little before they have attained to that level of technical skill which is necessary to a pianist of any claim to high excellence, was in Martin at its height and its noonday; while it really seemed sometimes to his master that he had been, perhaps in his cradle, perhaps as he bent his unwilling head over the crabbed intricacies of Demosthenes, somehow mysteriously initiated into the secrets of technique. Anyhow that facility, that art of first mastering and then concealing difficulties which to most pianists only comes, as it had come to himself, through months and years of unremitting toil, seemed to be natural to his pupil. Martin had only got to be told what to do, and if he was in an obliging humour he did it. The difficulties of execution simply did not seem to exist for him. Immensely struck as Karl Rusoff had been with his performance last summer at Lord Yorkshire's, he felt now that he had not then half fathomed the depth of his power, which lay pellucid like a great ocean cave full of changing lights and shadows, suffused to its depths with sunlight, and by its very clearness and brightness baffling the eye that sought to estimate its depths.

And his temperament—that one thing that can never be taught—Karl Rusoff knew he had never come across a temperament that, artistically speaking, approached it. It was indeed not less than perfect from that point of view, sensitive, impressionable, divinely susceptible to beauty, hating—(here largely was the personal charm of it to his master)—hating the second-rate, especially the skilful second-rate, with glorious intensity. At the thought Karl's rather grim face relaxed into a smile as he remembered how Martin had sat down to the piano the other day in a sudden burst of Handel-hatred, and with his ten fingers which sounded like twenty, and a strangely unmelodious voice which sounded like a crow and ranged from high falsetto treble to the note of kettledrums, had given a rendering of the Hailstone Chorus, so ludicrous, yet catching so unerringly the cheap tumult of that toy-storm in a tea-cup, that he himself had sat and laughed till his eyes were dim.

"And why," asked Martin dramatically in conclusion, "did that German spend his long and abandoned life in England? Because he knew, sir, he knew that in any other country he would have been kindly but firmly put over the border. Now shall I sing you the 'Hallelujah Chorus'?"

Besides this facility in technique, the power of perception of beauty, which in many of the finest minds requires years of delicate cultivation before it becomes at all mature or certain, was already present in Martin in apparent fulness of growth; it was already an instinct exerting and asserting itself, not through habit, but through intuition. It was so much the dominating ingredient in the composition known as Martin Challoner, that almost everything else might be considered as a mere by-product. His whole will, his whole energy was at its service. When once it called to him as it had called to him in his adoption of the Roman faith, it seemed he had to obey, and could not question. It was to him a law that he could not transgress.

But all this, the charm of which Karl Rusoff felt almost too keenly for his peace of mind, he knew to be extremely dangerous; and to him this exultant, beautiful mind was entrusted with all the responsibility that it entailed, to fashion, to train, to prune. With a true and honest modesty, he recognised how menial, so to speak, his work in regard to Martin was; but this did not lessen the responsibility. He was, to rate himself at the highest, the gardener who had to bring this exquisite plant into fulness of flower, to feed, to water, to cut, and above all to let air and sun, the great natural influences, have their way with it. He did not believe in forced growth, or in sheltered cultivation; as he had told Martin in the summer, every emotion, every pain and joy, so long as it was not sensual, was his proper food. The richer his experience was, the richer would his music be. Karl had already seen a first clear endorsement of his view in the circumstances attending Martin's secession to the Roman Church. He himself did not know with any exactitude of detail what had passed between him and his father; but though the painfulness of that had knocked Martin completely up for a time, what he himself had foreseen had come true, and he could hardly help inwardly rejoicing at even the cruelty of Mr. Challoner's attitude to his son, so great had been the gain to Martin artistically. He had suffered horribly, and was the better for it. Afterwards—the thing had taken place now more than two months ago—the elastic fibre of his youth had reasserted itself, and his exuberant health of body and mind had returned to their former vigour. The pain had passed, the gain remained.

Then to Karl's reverie there came the interruption he had been expecting: a quick step sounded outside, then a noise as of a large quantity of books being dropped in the passage, a loud and hollow groan, and after a short pause Martin, with half-a-dozen volumes of music, entered, flushed, vivid in face, muddy in boots.

"I am late," he said, "also I am sorry. But there was not a cab to be found. So I ran. I ran quicker than cabs. Oh, how hot I am!"

Karl's face lighted up as he saw him. He himself was unmarried and rather lonely in the world till this child of his old age had come to him, who should be, so he told himself, the crown of his life's work, and illuminate the dull world long after he himself was dead, with the melodious torch that he had helped to light.

"Are you late?" he said. "I have only just finished practising myself. My dear child, how hot you are! Let us have tea first. And are you dining out to-night? If not, have a chop with me here, and we can work a little afterwards as well. You have not been to me for a week."

"Yes, thanks, I should like that," said Martin. "I have been down at Chartries, as you know, for a couple of days."

He paused a moment, frowning at the fire.

"No, it was no good," he said. "My father would not see me. He even opposed Helen's coming to Uncle Rupert's while I was there. But she came."

"How is she?" asked Karl.

"Very well; and, what is so odd, extraordinarily happy:

happy in some steadily-shining way. Deep, broad, bright happiness, like sunlight. Now, how do you account for that? Away from Frank—she doesn't even write to him or hear from him—continuing to do all that she found so intolerable under hugely aggravated conditions—he not there—and yet awfully happy. Not that father has changed to her at all—he is very silent, very sad, very—well, sometimes very cross. And she feels his sadness, too—feels it as if it was her own——"

"Ah, you have hit it," said Karl; "that is why she is happy. It is what I have always told you—the fact of sympathy, whether it is with joy or pain, is what enriches and perfects; the fact of sympathy is what makes her happy. You are as happy, with the broad sunshine of happiness, even though a bitter wind whistles, when Isolde sinks lifeless by the body of Tristan, as when Siegfried hears the singing of the bird."

He paused a moment, looking at the fire, then turned to Martin.

"Ah, my dear lad," he said, "pray that you drink to the dregs any cup of sorrow or of joy that may be given you. Never shrink from pain; you will not become your best self without it. But by it and through it, and in no selfish or egoistic manner, you will fulfil yourself."

He rose from his chair and turned on switch after switch of electric light.

"It is like this," he said, feeling in his sudden desire for light some instinctive connection with what he was saying. "Open the doors, open the windows of your soul; let the sun in and the wind. And this is a music-lesson," he added laughingly. "Well, I have given a good many in my life, and should be pleased to know I never gave a worse one. Now, what have you done since I saw you last?"

Martin walked quickly over to the piano with a laugh.

"Listen," he said.

He played a few bars of very intricate phrase after the manner of the opening of a fugue. Then in the bass half the phrase was repeated, but it finished with something perfectly different—a third and a fourth and a fifth joined in, and before the "whole kennel was a-yelp," the original subject had passed through rapid gradations until it had become something totally different to what it began with, though still an incessant jabber of cognate phrases, never quite coherent, were somehow strung together and worked against each other by a miracle of ingenuity. Then the original subject was repeated with emphatic insistence, as if to call renewed attention to itself, but it was answered this time by a phrase that had nothing whatever to do with it; a third short melody, totally different from anything that had gone before or was to come after, ran its brief and ridiculous course, and then a perfect hodge-podge of reminiscences of all that had previously occurred, handled with extraordinary dexterity, made the brain positively reel and swim. Finally, a huge bravura passage, as much decked out with ribands and lace as a fashionable woman at a party, brought this insane composition, which taxed even Martin's fingers, to a totally unexpected close.

Karl Rusoff had listened at first with sheer uncomprehending bewilderment, unable, since indeed there was neither head nor tail nor body to it, to make anything whatever out of it, and for a moment he wondered if Martin was merely playing the fool. But as he looked at his face bent over the piano, and saw even his fingers nearly in difficulties, a sudden light struck him, and he began to smile. And before the end was reached he sat shaking in his chair with hopeless laughter.

"Ah, you wicked boy!" he said. "Why, even our dear Lady Sunningdale would recognise herself."

Martin pushed his plume out of his left eye, and laughed.

"That's the joy of it," he said. "She did recognise it. About half-way through she said, 'Why, that's me!' You know you told me to do that, to take anything, the east wind, or a London fog, or a friend, and make music of it."

"Play it once more, if you will," said Karl, "and then to work. Not that that is not work. There is a great deal of work in that; also I perceive with secret satisfaction that you do not find it easy to play. But the bravura is rather unkind. She is never quite like that."

"Ah! the bravura is only her clothes," said Martin, preparing to begin again. "She even told me which hat she had on. It is the one she describes as a covey of birds of paradise which have been out all night in a thunderstorm, sitting on a tomato-salad."

Again Karl sat and listened to the torrent of fragments and currents of interrupted thoughts. Heard for the second time, it seemed to him even a more brilliantly constructed absence of construction than before, an anomalous farrago which could only have been attained by a really scholarly and studious disregard of all rules; no one who had not the rules at his finger-tips could have broken them so accurately. It was a gorgeous parody of musical grammar in exactly the mode in which Lady Sunningdale's conversation was a brilliant parody of speech, full of disconnected wit, and lit from end to end with humour, but as jerky as the antics of a monkey, as incapable of sustained flight in any one direction as a broken-winged bird, a glorious extravagance.

Karl had left his seat and stood near the piano as the bravura passage began. This time it seemed to present no difficulty to Martin, though his unerring hands were hardly more than a mist over the keys. And Karl felt a sudden spasm of jealousy of his pupil, as a huge cascade of tenths and octaves streamed out of Martin's fingers.

"Yes, indeed, the bravura is not easy," he remarked when Martin had finished, "and I think you played it without a mistake, did you not? Is it *quite* easy to play tenths like that?"

Martin laughed.

"I find I've got not to think of anything else," he said. "Will that do for my composition for the week?"

Karl laughed.

"Yes, very well indeed," he said. "It has lots of humour, and humour in music is rather rare. But don't cultivate it, or some day you will find yourself in the position of a man who can't help making puns. A dismal fate! Now let us leave it (it is admirable) and get to work. I think I told you to study the last of the 'Novelletten.' Play it, please."

This time, however, there was no laughter and no approbation. Karl looked rather formidable.

"It won't do; it won't do at all," he said. "You have the notes, but that is absolutely all. It is perfectly empty and dead; a pianola would do as well. What's the matter? Can't you read anything into it?"

Martin shrugged his shoulders.

"I know it's all wrong," he said. "But I can't make anything of it. It's stodgy."

Karl's eyes glared rather dangerously from behind his glasses.

"Oh, stodgy is it?" he said slowly. "Schumann is stodgy! That is news to me. I must try to remember that."

Martin looked sideways at his master, but Karl's face did not relax.

"Stodgy!" he repeated. "I know where the stodginess comes in. Ah! you are either idiotic or you have taken no trouble about it. Because you have found that the mere execution was not difficult to you, you have not troubled to get at the music. I gave you music to learn, and you have brought me back notes. Do not bring a piece to me like that again. If I give you a thing to learn, I do so for some reason. Get up, please."

Karl paused a moment, summoning to his aid all that he knew, all he had ever learned, to give cunning to his fingers and perception to his brain. Never perhaps in his life had he played with more fire, with more eagerness to put into the music all that was his to put there, and that in order to charm no crowded hall packed from floor to ceiling, but to show just one pupil the difference between playing the music and playing the notes.

Martin had left the music-stool in what may be called dignified silence, and was standing by the fire, but before long Karl saw him out of the corner of an eye (he could spare him neither thought nor look) steal back towards the piano, and though he could not look directly at his face he knew what was there: those wide-open black eyes, finely chiselled nostrils swelling and sinking with his quickened breath, mouth a little open, and the whole vivid brain that informed the face lost, absorbed.

He came to the end and sat silent.

"Is that there?" asked Martin in a half breathless whisper. "Is that really all there?"

Karl looked up. Martin's face was exactly as he had known it would be. But the first mood of the artist was of humility.

"I played wrong notes," he said. "Half-a-dozen at least."

"Oh, more than that," said Martin. "But what does that matter? You played it. My God, what a fool I have been! There I sat day after day, and never saw the music."

Karl Rusoff got up. It had been a very good music-lesson.

"It isn't 'stodgy,'" he said. "It isn't really. Do you now see one thing out of a hundred perhaps that it means? You have got to be the critic of the music you play; you have to interpret it. But out of all the ways of playing that, out of all that can be seen in it, you saw nothing; your rendering was absolutely without meaning or colour. To play needs all you are; you gave that fingers only. If I want you to practise fingers only, I will tell you so, and give you a finger exercise or Diabelli. Otherwise you may take it for granted that when your fingers are perfect your work begins. But to play—ah!—you have to burn before you play."

Martin still hung over the piano.

"And I thought it stodgy!" he repeated, looking shy and sideways at Karl's great grey head.

"Well, you won't again," said he. "Will you try it again now?"

"No, how can I?" said Martin. "I've got to begin it all over again."

"Then there was a piece of Bach. Play that. And now read nothing into it except the simplicity of a child. Just the notes, the more simply the better. Wait a moment, Martin. I want to enjoy it. Let me sit down."

Martin waited, and then began one of the "Suites Anglaises," and like a breath of fresh air in a stuffy room, or like a cloudless dawn with the singing of birds after a night of storm and thunder, the exquisite melody flowed from his fingers, precise, youthful and joyous. There was no introspection here, no moods of a troubled soul: no doubts or questioning: it sang as a thrush sings, changed and returned on itself, danced in a gavotte, moved slowly in a minuet, and romped through a Bourrée like a child.

At the end Martin laughed suddenly.

"Oh, how good!" he cried. "Did you know that Bach wrote that for me?" he asked, turning to Karl.

"Yes, I thought he must have," said Karl. "And with the command that you were to play it to me. You played that very well; all your fingers were of one weight. How did you learn that?"

Martin raised his eyebrows.

"Why, it would spoil it, would it not, to play it any other way?" he asked.

"Certainly it would."

Then he got up quickly.

"Oh, Martin, you child!" he said. "Did I speak too roughly about the Schumann?"

"You did rather," he said. "But I deserved to have my ears boxed."

The two dined alone, and held heated arguments,

not like master and pupil but like two students who worked side by side, Karl as often as not deferring to the other, Martin as often as not blandly disagreeing with Karl.

"Now can you pronounce, for instance," he asked, "that that 'Novelette' is to be played with those sweatings and groanings, the mere notes being of no use, whereas Bach is to be played with notes only?"

Karl gazed at him in silence.

"You impertinent infant!" he said. "What else do you propose? To play the Schumann as you played it? And the Bach as I played the Schumann?"

"That would sound extremely funny," remarked Martin. "No, I don't say you are not right, but how do you know you are right?"

"Because Bach wrote for the spinnet," said Karl. "Have you ever tried to play Schumann on a spinnet? It sounds exactly as you made it sound just now. A deplorable performance, my poor boy."

"You have told me that. Don't rub it in so. I shall play it very well to-morrow."

"Or next year," said Karl, still grim but inwardly full of laughter. "By the way, there was no 'dog' motive in the Lady Sunningdale composition."

"You can't have been attending," said Martin. "Suez Canal came in twice, and Sahara three times with shrill barks. Yes, please, another cutlet."

Karl watched him eat it: the process took about five seconds.

"You didn't taste that," he remarked.

"No, it was needed elsewhere," said Martin. "But I'm sure it was very good."

Karl lingered over the bouquet of his Burgundy.

"It is a strange thing," he said, "that mankind are so gross as to confuse the sense of taste with greediness. No, my dear boy, I am not at this moment attacking you. But there is no organ, even that of the ear, in this wonderful body of ours, so fine as that of taste. Yet to most people the sight of a man deeply appreciating his dinner conveys a feeling of greediness. But I always respect such a man. He has a sense more than most people."

"But isn't it greedy?" asked Martin.

Karl became deeply impressive.

"It is no more greedy," he said, "to catch the flavour of an olive or an oyster, than to catch the tone of a 'cello."

"But a hundred oysters?" said Martin.

"Ah, that would be like encoring a song in an opera—a most detestable habit—and hearing it over and over again. No artist desires that. Fancy hearing Wotan's 'Abschied' twice. That would be greedy. The art of dining, like most arts, is frightfully neglected in England."

Martin laughed.

"I have been here, I suppose, a dozen times," he said, "and every time you give me some surprise. I had no idea you gave two thoughts as to what you ate."

"That was hasty of you. True, of all the senses I put the ear first. That is personal predilection. But all the senses really are equal; there is no shadow of reason for supposing that one is more elevated than another. True, some can be more easily misused than others, taste more particularly. But all are subtle gateways to the soul."

They had finished dinner and Karl pushed back his chair.

"Take an instance," he said. "Take incense. Does not that smell excite and inspire the devotional sense? Does not the smell of frangipanni—an unendurable odour—suggest a sort of hothouse sensualism? Does not the smell of a frosty November morning bring the sense of cleanness into the very marrow of your bones?"

Martin sniffed experimentally.

"Ah, I know that!" he said. "And the leaves on the beech-trees are red, and the grass underfoot a little crisp with frost. Oh, how good! But what then?"

Karl was watching him closely. It was his conscious object now and always to make Martin think, to excite anything in him that could touch his sense for beauty. He had found that this half-serious, half-flippant method was the easiest means of approach. For Martin was but a boy: discussions in an earnest, conscious German spirit both bored and alarmed him. This fact, had his father grasped it, might in years past have helped matters.

"Why, everything," he replied. "Each sense can be expressed in terms of another. Take magenta in colour: it is frangipanni in smell, in sound it is—what shall we say?—an Anglican chant of some sort, in taste it is the vague brown sauce in which a bad cook hides his horrors."

Martin laughed again, with the keen pleasure of youth in all things experimental.

"Yes, that is true," he said. "How do you go on? Take a fine colour, vermilion."

"The blind man said it must be like the sound of a trumpet," said Karl, "and the blind man, at that moment, saw. Brandy also for taste is red. So is ammonia: a pistol to your nostrils."

Martin dabbed his cigarette onto his dessert plate.

"Yes, yes," he cried, "and C major is red. And F sharp is blue, electric blue, like the grotto at Capri——"

He stopped suddenly.

"Am I talking nonsense?" he asked. "If so, it is your fault. You encouraged me. You meant to. And what do you mean me to get from it?"

Karl turned directly towards him.

"I mean you to think," he said. "To frame your life wholly for beauty, in whatever form you see it. It is everywhere, be assured of that; and if your eye sees it, store it up like a honey bee, and bring it home. If your mouth feels it, bring it home. If you smell the autumn morning, bring that home too. It all makes music."

He pushed his plate aside, and leaned forward towards Martin.

"All is food for you," he said. "It is only in that way, by harvesting every grain of corn you see, that you can be great. A lot of harvesting is done unconsciously. Supplement that by conscious harvesting. You may learn perfectly all the harmony and counterpoint that can be learned, you may learn to play things impossible, but all that is no good by itself. You can already play—I am not flattering you, but the reverse—if you practise a little, all the printed music ever written, as far as notes go. That is no good either. But—if I had not seen this when first I heard you play, I should never have wasted ten minutes of my time on you—you can do more than that. You can, if you are very alert, quite untiring, very critical, and always ready to catch beauty in whatever form it may present itself, you can do more than this. At least I believe so."

He got up from his seat, and leaned his hands on the boy's shoulders as he sat by the table.

"Ah, Martin, don't disappoint me," he said, "or, being old, I shall die of it. Drink from every spring but one, and drink deep."

Martin turned in his chair and faced him.

"Do I know what spring you mean?" he asked. "Love?"

Karl looked at him with a sort of wonder.

"No, I did not mean that," he said.

He drew a long breath.

"My God, if that had been granted to me," he said, "I too might have been great. But I never fell in love. Oh! I am successful; I know, I understand. I am the only person perhaps who does know what is missing in me. It is that. But missing that, I never, no not once, parodied what I did not know. Parody, parody!" he repeated.

Martin looked at him with that direct lucid gaze Karl knew so well, level beneath the straight line of his eyebrows. His smooth brown cheeks were a little flushed with some emotion he could not have put a name to. Slight injury was there, that Karl could possibly have supposed him bestial, the rest was clean modesty.

"I am not beastly," he said, "if you mean that."

"I did mean that," he said. "And I beg your pardon."

Martin stood up.

"I think you had no right to suppose that," he said.

"No, I had none. I did not suppose it. I warned you, though."

A tenderness such as he had never known rose like a blush into his old bones, tenderness for this supreme talent that had been placed in his hands.

"I only warned you," he said. "I looked for burglars under your bed, just because—because it is a boy like you that this stupid world tries to spoil. Aye, and it will try to spoil you. Women will make love to you. They will fall in love with you, too."

Again he paused.

"Things will be made poisonously pleasant for you," he said. "You can without effort capture brilliant success. But remember all that you get without effort is not, from the point of view of art, conceivably worth anything. Remember also that nothing fine ever grew out of what is horrible. More than that, what is horrible sterilises the soil—that soil is you, you will never get any more if you spoil it or let it get sour or rancid. Horror gets rooted there, it devours all that might have been good, all that might have been of the best."

There was a long silence. Then Karl stepped back and rang the bell. To Martin the silvery tinkle sounded remote. He certainly was thinking now.

"Well, I have done," said Karl. "Excuse the—the Nonconformist conscience."

Martin got up.

"I don't see how one can care—really care—for music and live grossly," he said. "Yet people appear to manage it. And mawkishness makes me feel sick," he added with apparent irrelevance.

But Karl understood.

"Somebody has been trying to pet him," he thought to himself.

They went upstairs to the music-room, and Martin stood before the fire a few moments smoking in silence.

"I like this room," he said. "It makes me feel clean like the November morning. I say, how is it that so many people, men and women alike, only think about one subject? Surely it is extraordinarily stupid of them, when there are so many jolly things in the world."

"Ah, if the world was not full of extraordinarily stupid people," remarked Karl, "it would be an enchanting place."

"Oh, it's enchanting as it is," cried Martin, throwing off his pre-occupation. "May I begin again at once? I want to get through a lot of work to-night. Heavens, there's a barrel-organ playing 'Cavalleria.' Frank is going to introduce a Bill next Session, he says, putting 'Cavalleria' in public on the same footing as obscene language in public. He says it comes to the same thing."

Stella Plympton about this time was giving a certain amount of anxiety to her parents. The amount, it is true, was not very great, because her father was a happily constituted man who was really incapable of feeling great anxiety except about large sums of money. Consequently, since the extremely large sums of money, all of which he had made, were most admirably invested, his life was fairly free from care. His wife also was quite as fortunate; her complexion was the only thing capable of moving her really deeply, and as she had lately found a new *masseuse* who was quite wonderful, and obliterated lines with the same soft completeness with which bread-crumb removes the marks of lead-pencil, she also for the present stood outside the zone of serious trouble.

Between them they occupied just now the apex of social as well as most other successes in London, and were a very typically modern couple. Sir Reginald Plympton had in early life invented an oilcloth of so eminent an excellence that in its manufacture and exploitation he had been too busy to really master the English aspirate, which still bothered him. But to make up for this he had carefully cultivated his aspirations, and had (entirely owing to oilcloth) while not yet sixty amassed a colossal fortune, married the daughter of an impecunious Duke, won the Derby, and now stood perched on the topmost rung of the ladder of English society. He had a yacht, which never went for long cruises, but always anchored for the night in some harbour. Being a bad sailor he left it, if there was a chance of bad weather, before it weighed anchor in the morning, and joined it again on the ensuing evening. Similarly he sat in his wife's opera-box during intervals between the acts, and left his place on the rising of the curtain. He was already a baronet and an M.P., and his peer's coronet, so to speak, was now being lined.

Yet care, though only like a little draught, just stirred the warm air of Lady Monica's drawing-room, and made the palm-leaves rattle. She had often talked the matter over with her husband, who had no very practical suggestion to make. He would stand before her, very square and squat, with his hands in his pockets, rattling money in the one and keys in the other, and say:

"Well, my lady, you give 'er a good talkin' to. Tell 'er to be a good girl, and be sensible. And now I must be off."

For the fact was that Stella was now nearly twenty-three; she had refused several very suitable offers, and her mother, extremely anxious, as all good mothers should be, to get her married, had lately begun to be afraid that she was "being silly." This, in her vocabulary, meant that Stella was in love with somebody (Lady Monica thought she knew with whom), and was not clever enough to make him propose to her. What added enormity to her "silliness" was the fact that he was extremely eligible. Lady Monica had no sympathy with this sort of thing; she had never been silly herself, and her own sentimental history had been that some twenty-five years ago she had wooed, proposed to, and wedded her Reginald, without any fuss whatever, or any delay. She was a woman with a great deal of hard, useful commonsense; she always knew exactly what she wanted, and almost always got it. Her only weakness in fact (with the discovery of the new *masseuse,* her complexion had become a positive source of strength) was for feeble flirtations with young men of the age which she herself wished to look. These never came to anything at all, and when the young man in question married somebody perfectly different, she told all her friends that she had made him. She had during the last week or two, since the Session had brought them to London, done a little vicarious love-making to Martin on Stella's account, and enjoyed it on her own. She was a perfectly honest woman, and only played with fire as a child plays with matches, lighting them and blowing them out, and she never really set fire to herself, and quite certainly never even scorched anybody else.

But anxiety, like a draught, had reached her with regard to Stella's future, and the next evening when Lady Sunningdale happened to be giving a menagerie-party, she determined to have a few words with her, for she was regarded as a sort of book of reference with regard to the Twins. The menagerie-party was so called because for a week beforehand Lady Sunningdale drove about London a good deal and screamed an invitation to everybody she saw in the streets. The lions only were fed; the meaner animals and those lions only observed too late to ask to dinner came in afterwards.

Lady Monica and Stella belonged to this second category, and Lady Sunningdale hailed them with effusion.

"Dearest Monica, so glad to see you!" she cried. "All sorts of people are here, whom I'm sure I don't know by sight, and I've just revoked at Bridge (doubled no trumps too, isn't it too dreadful?) and Suez Canal tried to bite the Prime Minister. Wasn't it naughty, but you see Suez is a Radical, though he shouldn't bring politics into private life. Stella, I haven't seen you for years. Yes; Martin's going to play, of course. Have you heard his tune which imitates me talking in a very large hat? Simply heavenly; exactly like. Even Sunningdale awoke the other evening when he played it, and asked me what I was saying. How are you, Frank? No sign of relenting on the part of the obdurate father? How dreadful! Yes. Dearest Monica, how well you are looking, and how young! ('New *masseuse,*' she thought to herself, 'I must worm it out.') Do let us go and sit down; I'm sure everybody has come. Oh, there is the Spanish Ambassador; he killed his own father you know, shot him dead on the staircase, thinking he was a burglar, and came into all that immense property at the age of nineteen! How picturesque, was it not, and such a very Spanish thing to do! Such a good shot, too! How are you, Señor? Yes, they are playing Bridge in the next room. And they say there is sure to be a dissolution in the Autumn."

Lady Sunningdale poured out this spate of useful information in her usual manner, addressing her remarks indiscriminately to anyone who happened to be near, and Lady Monica waited till the flood showed some sign of abating. She had a vague contempt for Lady Sunningdale's "methods," considering that she diffused herself too much: she never caught hold of anything, and held tight till everybody else who wanted it let go from sheer fatigue, which was a favourite method of her own. On the other hand, Lady Sunningdale certainly managed to pick up a great many bright objects as she went along, even though she did drop them again almost immediately.

"Do come away and talk to me, Violet," said Lady Monica, when for a moment there was silence. "I came here entirely to have a confabulation with you."

"Yes, dear, by all means. I have heard nothing interesting for weeks except the things I've made up, and told in confidence to somebody, which have eventually come round to me again, also in confidence. What's it all about?"

As soon as they had found a corner, Lady Monica, as her custom was, went quite straight to the point.

"It's about Stella," she said. "Violet, I am afraid Stella is being silly."

"How, dear? Stella always seems to me so sensible. Such a lovely neck too, quite like yours! Look, there is poor Harry Bentham. A lion bit his arm off, or was it South Africa?"

Lady Sunningdale cast a roving eye in his direction, kissed the tips of her fingers, and motioned him not to come to her. Lady Monica waited without the least impatience till she had quite finished. Then she went on, exactly where she had left off.

"Well, it's your dreadfully fascinating Martin Challoner," she said, "and I'm sure I don't wonder. My dear, really such terribly attractive people ought to be shut up, not allowed to run about loose. They do too much damage."

"Well, dear, Stella is only like all the rest of us," said Lady Sunningdale. "You remember how we all ran after the Twins last summer."

"I know; we all got quite out of breath. But Stella is running still. Now do you think, you know him so well, that he gives two thoughts to her? They are great friends, they are often together, but if it is all to come to nothing, I shall stop it at once. Stella has no time to waste."

Lady Sunningdale considered this a moment. She knew all about Monica's little flirtations with Martin, so also did he, and had imitated her for Lady Sunningdale's benefit with deadly accuracy. But she was too good-natured to spoil sport just because Stella's mother had been a shade too sprightly for her years. Besides, she meant to say a word or two about that later on, a word that would rankle afterwards.

"My dear, I can't really tell whether Martin ever thinks about her or not," she said. "He is so extraordinary; he is simply a boy yet in many ways, and he plays at life as a boy plays at some absurd game, absorbed in it, but still considering it a game. Then suddenly he goes and does something deadly serious, like joining the Roman Church. Practically also you must remember that he thinks almost entirely about one thing, his music. That child sits down, and plays with the experience and the feeling and the fingers which, as Karl Rusoff says, have never yet been known to exist in a boy. He is like radium, something quite new. We've got to learn about it before we can say what it will do in given circumstances. It burns, and it is unconsumed. So like Martin! But Karl says he is changing, growing up. I can't help feeling it's rather a pity. Yes. Of course he can't be a bachelor all his life, that is impermissible, but Karl always says, 'I implore you to leave him alone: don't force him, don't even suggest things to him. He will find his way so long as nobody shows it him.' Karl is devoted to him, just like a beautiful old hen in spectacles with one chicken."

But Lady Monica had not the smallest intention of talking about Karl, and led the conversation firmly back.

"Well, Violet, will you try to find out?" she asked.

Lady Sunningdale's eye and attention wandered.

"Ah, there is Sunningdale," she said. "Does he not look lost? He always looks like that at a menagerie. Yes, I will try to sound Martin, if you like. I must make him confide in me somehow and be rather tender, and he will probably tell me, though he will certainly imitate me and my tenderness afterwards. He imitates people who take an interest in him—that is his phrase—too beautifully. I roared—" Lady Sunningdale cast a quick sideways glance at her friend—"simply roared at some imitation he gave the other day of a somewhat elderly woman who took an interest in him. Yes. Poor Suez Canal! He loves parties, but one can't let him bite everybody indiscriminately. Let us come back, dear Monica, and make the Twin play. There he is sitting with Stella. He asked me particularly if she was coming. They are probably talking about golf or something dreadful. Stella is devoted to it, is she not? Yes. That's the game where you make runs, is it not? I shall have to sound Martin very carefully. He is so quick. Sunningdale, please take Martin firmly by the arm, and if he tries to bite, by the scruff of the neck, and put him down at the piano. No, dear Monica, you can tell nothing by his face. He always looks absorbed and excited like that. If he was talking to you he would look just the same."

That also was premeditated and vicious, just in case poor Monica's little love-making, which Martin had imitated so divinely, had not been wholly vicarious. If it had, her remark would pass unnoticed, if it had not—but there was no need to consider whether it had or not, for poor Monica had turned quite red at the mention of Martin's imitation of the elderly woman who took an interest in him.

CHAPTER XI.

MARTIN had been among the lions who were fed to-night at Lady Sunningdale's, and had eaten of rich and slightly indeterminate food, for his hostess's vagueness and volubility, like Karl's love of form, found expression even in the dinner. Afterwards he had taken up a strategic position near the head of the stairs, when the meaner animals or belated lions began to arrive, in order to watch and wait for Stella's entrance. Then as soon as her mother and Lady Sunningdale had retired into their corner, he had annexed her—with her complete assent—and plunged into discussions about affairs not in the least private. Had her mother overheard, she would, with her strong practical commonsense, have ordered the conversation to cease at once, so wanting in the right sort of intimacy would she have found it. And in so doing she would have made one of those mistakes which are so often and so inevitably committed by people of great commonsense but no imagination, who cannot allow for the possible presence of romance in pursuits which they themselves consider prosaic. Had Martin been talking to her daughter about music, she would have considered that sufficiently promising to allow developments, for that was a thing very real to him; his heart spoke. As it was, she would have considered that the conversation held not a germ of that disease of which she longed that Martin should sicken.

Lady Sunningdale, far less superficial really than the other, and knowing that almost everything under the sun was rich with childish romance in Martin's eyes, had hazarded the suggestion that they were talking about golf. This was practically correct, because they were talking about skating, and the two to her were indistinguishable—she supposed you got runs at each—being objectless exercises for the body. The moment you hunted or shot or played any game you entered that bracket. All these things were of the same *genre,* and quite unintelligible.

"But I can't get my shoulders round," said Stella. "It is no earthly use telling me that I must. They won't go. Can you understand the meaning of those three simple words, or shall I try to express it differently? And if I try to make them get round, I fall down."

Martin frowned.

"Stella, you are really stupid about it," he said—they had long ago fallen into Christian names—"for the hundredth time you have to consider your foot as fixed. Then pivot round, head first—then——"

Stella nodded.

"Yes, I understand that," she said. "It is always head first with me. On the ice."

"You're not being serious," said Martin, "and if you can't be serious about a game you can't be serious about anything. That is a universal truth. I discovered it. What do you suppose matters to me most in my life? Music? Not at all. Get along with you, you silly thing! But, oh, if anyone would teach me to do back brackets not rather clean, but quite clean. I dreamed I did one once, and I awoke sobbing loudly from sheer happiness. I would sign a pledge never to touch tobacco or a piano again, if I could do that. That's my real state of mind. Now, will you skate to-morrow at Prince's? I can be there at ten for an hour."

"Considering I am always there at half-past nine," remarked Stella, "I don't think you need ask. And yet you say I am not serious! Oh, Martin, why is it that one really only wants to do the things one can't do?"

"You can if you want enough," said he. "The deuce is that one can't always want enough."

"I don't believe that," said she promptly. (Lady Monica would have stayed her devastating hand if she had heard this.) "I want lots of things as much as I possibly can."

"But perhaps even that isn't enough. What, for instance?"

Stella could not help a momentary lifting of her eyes to his.

"Why, to skate, silly," she said. "Yes, I'll be there by ten, and so do be punctual. I will consider my foot whatever you wish, and I'll fall down as often as you think necessary. But don't be unkind at once when you pick me up, and tell me I was too much on my heel, or anything of that sort. Wait till the first agony is over. I attend best when the pain is beginning to pass off."

"Well, I only tell you to save trouble in the future," said he.

"I know, but give me a moment. Do you care about the future much, by the way? I don't. Give me the immediate present. To think much about the future is a sign of age; no one begins to care about the future until he is too old to have any. Besides, it implies that the present has ceased to be absorbing."

Martin pondered this.

"Oh no, I don't think that is so at all," he said.

Stella laughed.

"You never by any chance agree with a word I say," she remarked.

"Well, you haven't agreed with me since August," he said. "I made a note of it. But that is why we have no stupid pauses. All conversation runs dry in two minutes if one agrees with the other person. But what you say about age really isn't so. Look at Karl Rusoff or Lady Sunningdale. They both live intensely in the present."

"Ah, you are shallow," she said. "Years have got nothing whatever to do with age. That is the *most* superficial view. People of ninety die young, people of twenty die of senile decay."

Martin stretched his trouser over his crossed knee.

"I am a hundred and eleven," he said, "and whiles——don't you hate the Scotch—and whiles I am about twelve in an Eton collar."

"Yes, loathe them, laddie! Hoots! That is what is so maddening about you. Half the time I think I am talking to my great-uncle, and the rest of it to my little nephew up from the country."

"Is he a nice boy?" asked Martin. "Or do you like your great-uncle best?"

"I don't like either at all, thank you. You are always being far too wise or far too young. As a man of a hundred, how can you play silly games with such enthusiasm? And as a boy of twelve, how can you play the piano as you do?"

"It is because I am so extremely gifted," said Martin, so gravely and naturally that for an appreciable moment she stared.

"Ah! Don't you find it an awful bore?" she asked.

"Dreadful. I can't really take any pleasure in anything, owing to the sense of responsibility which my talents bring to me."

Stella broke down and laughed; at gravity he always beat her completely. At which period in their conversation Lord Sunningdale did as he was ordered, and taking him firmly by the arm, led him to the piano.

Karl was always most assiduous in his attendance at houses where Martin played, and he was here to-night. His object was certainly not to flatter or encourage his pupil; for often and often, when Martin had played in his presence the night before, he found but a growling reception waiting for him at his next lesson.

"You played well enough for them," Karl would say; "I grant you that. Any bungling would do for them. But to play 'well enough for them' is damnation."

"But it *did,*" Martin would argue. "I did not want to play at all, but one can't say no; at least, I can't. I was not playing for you."

"Then you should not have played at all. If you play often enough in a second-rate manner, you will soon become second-rate."

But to-night Martin never suggested the second-rate even to his exacting master. In a sort of boyish protestation at the strictures he had undergone last night concerning the last of the "Noveletten," he played it again now. Certainly to-night there was no note of stodginess there; the varied, crisp, masterful moods of the music rang extraordinarily true. Half-way through Karl turned to Lady Sunningdale, who was sitting next him:

"How has he spent his day?" he asked suddenly.

"Skating, I think; he skated all morning and was late for lunch, and he went back to Prince's afterwards. He is terribly idle, is he not? Pray don't interrupt,

Monsieur Rusoff; I never can feel as if I hear a note at all unless I hear them all. Who said that? You, I think. So true! And have you heard his piece on me? He must play it. Delicious this is, isn't it? I learned it when I was a child. Tum-te-tum. There is the tune again."

"But with whom did he skate, my dear lady?" asked Karl. There had been a good many notes missed by now.

Lady Sunningdale gasped.

"Oh, Monsieur Rusoff, how clever of you!" she said. "You are really clairvoyant. So is my maid, the one like a murderess. Do you know her? No, how should you? Martin was skating with Stella Plympton. And that is important, is it? Don't tell her mother. She is such a fool, and also she has been trying to pump me. You see, it was I who brought them together. So suitable! I feel dreadfully responsible——"

At this point the "Novelette" ended, and Lady Sunningdale clapped her hands in a perfunctory manner.

"Too heavenly, monster!" she said. "Now play Tum-te-tum. Yes, that one. And is he really going to marry her?" she continued to Karl. "I love being pumped, if I know it. Dear Monica, she pumps like a fire-engine. There is no possibility of mistake. Now, while he is playing this, do tell me all you know."

"My dear lady, you are building on no foundation," said Karl. "All I know is that he played that to me last night, and played it abominably. To-night he has played it—well, you have heard. And psychologically, I should like to know what has occurred in the interval."

"Was his playing of it just now very wonderful?" she asked.

"Yes; one might venture to say that. And as he has been skating all day, presumably he has not thought much about it. His thinking perhaps has been done for him. And who is Stella Plympton? Wife or maid?"

Lady Sunningdale gave a little shriek of laughter. Really people who lived out of the world were much more amusing than those who lived in it. Those who lived in it, it is true, always believed the worst in the absence of definite knowledge; the others, however, made far more startling suggestions.

"Next but two on your right," she whispered. "Dear Monica will have a fit if Stella turns out to be already married."

Karl's eyes wandered slowly to the right, looking pointedly at many things first, at the cornice of the ceiling, at Martin's profile, at the slumber of Lord Sunningdale. Then they swept quickly by Stella.

She sat there absorbed and radiant, her face flushed with some secret delicate joy as she watched and listened, hardly knowing whether eyes or ears demanded her attention most. Certainly the music and the musician between them held her in a spell.

"She is looking quite her best," whispered Lady Sunningdale. "How interesting! They have millions, you know—oil-cake, or was it oil-cloth? Oil something anyhow, which sounds so rich, and she is the only child. The father is quite impossible, not an 'h,' though everyone crowds there. One always does if there are millions. So vulgar of one! Dear Monica, we were almost brought up together."

Karl turned round to her.

"Dear Lady Sunningdale," he said, "you are really quite premature if you build anything on what I have said. He played admirably to-night what he played abominably last night. That is absolutely all I know. I should be so sorry if I had suggested anything to you which proved to be without any sort of foundation."

There certainly seemed to be some new power in Martin's playing to-night; but new power had constantly shown itself there during the last month or two, for as Karl said he had been growing. To-night, however, he was conscious of it himself, and even as he played he knew that fresh light of some kind, some fresh spring of inspiration, was his. His hand and his brain were too busy as he played to let him be more than conscious of it; where it came from, what it was, he could not guess this moment; but as he struck the last chords the tension relaxed, and he knew. Then, looking up, he saw Stella sitting near him, leaning forward, her beautiful mouth a little open. That glorious white column of her neck supported her head like the stem of a flower, no garden flower, but something wonderful and wild. There were rows of faces behind her, to each side of her; she was one in a crowd only, but as his eyes caught her gaze, the crowd fell away, became misty to him, vanished as a breath vanishes in a frosty air; and she only, that one face, bending a little towards him, remained.

For a long moment their eyes dwelt on each other; neither smiled, for the occasion was too grave for that, and they two, for all they knew, were alone, in Paradise or in the desert; it was all one. The gay crowd, the applause that merged into a crescendo of renewed conversation, lights, glitter, men and women, were for that one moment obliterated; for in his soul love had leaped to birth, no puny weakling, prematurely warped and disfigured by evil practices and parodies of itself, but clean and full-grown it sprang towards her, knowing, seeing that its welcome was already assured. Then the real world, so strangely unreal in comparison to that world in which for a moment their souls had mingled and embraced, reeled into existence again, and Martin rose from the piano, for Stella had risen also, and had turned to some phantom on her right that appeared to speak to her.

Lady Sunningdale beckoned and screamed to him.

"Martin," she cried, "you are too deevey! Monsieur Rusoff is really almost—didn't you say almost?—satisfied with the way you played that. And you learned all that exquisite thing—I used to play it years ago—while you were skating to-day, because he says you played it too abominably last night. Really, if I thought I could play it like that to-morrow evening, I would go and skate all day. Now don't waste time, but play something more instantly."

"Oh, please, Lady Sunningdale, I would rather not," said he. "I really don't think I could play any more to-night. I really am—I don't know what—tired."

Lady Sunningdale looked at his brilliant, vigorous face.

"Martin, I don't believe you will ever learn to tell a decent, passable lie," she said. "Why not tell me you have got cancer? Oh, there's Suez Canal come back. Naughty! Monsieur Rusoff, won't you tell him that he must? Just a scale or two. I adore scales, so satisfactory, are they not?—so expected—as if it was a music lesson. No? How tiresome of you!"

Karl laid his hand on Martin's arm.

"No, my dear lady," he said. "He is never to play except when he wants to. But if you really want a little more music, and I——"

"Ah! but how enchanting of you! Monsieur Rusoff is going to play. Surely, dear Monica, you will wait? You are not going yet?"

"Desolated, Violet; but Stella says she feels a little faint. The hot room, I suppose; she is waiting for me outside. How deliciously you play, Mr. Challoner! I suppose you practise a great deal? Won't you come some day and——"

She broke off, for Martin had simply turned his back on her, and was firmly edging his way through the crowd to the door. Then Lady Monica's maternal instinct positively leaped to a conclusion, and Martin's rudeness was completely forgiven.

"But I can't resist waiting to hear Monsieur Rusoff," she said. "I thought he never played at private houses. How clever of you, dear Violet! I wonder if you could get him to play for me. Stella will sit down and wait for me, no doubt."

But before Karl struck the first chord, Martin had won (not to say pushed) his way through the hushed crowd, and found Stella sitting outside in the other drawing-room. Everyone had flocked in to hear the music, and they were alone.

His foot was noiseless on the thick carpet, and he was but a yard or two from her when she raised her eyes and saw him. Then, with a little choking cry, only half articulate, he came close to her. All the excitement and fire in which his life was passed was cold ashes compared to this moment, and his heart thumped riotously against his chest. Twice he tried to speak, but his trembling lips would not form the words, and she waited, her eyes still fixed on his. Then suddenly he threw his arms out.

"It is no good trying," he said. "But I love you; I—I love you."

Oh, the clumsy, bald statement! But Life and Death meant less than that word.

"Oh, Martin," she said, "I have waited; I—I don't know what I am saying."

"Waited?" he asked, and his eyes glowed like hot coals.

Then he laughed.

"And you never told me," he said. "If it was not you, I should never forgive you. And if it was not you, I should not care."

"Isn't that nonsense?" she asked.

"Yes, probably. Who cares? Stella! Oh, my star!"

He flung his arms round her.

"My star, my star!" he cried again.

For one moment she could not but yield to him.

"Yes, yes," she whispered; "but Martin, Martin," and her mouth wreathed into laughter, "it is an evening party. You must not; you must not."

He paused like a man dying of drought from whose lips the cup of water had been taken away.

"Party!" he cried, "what party? It is you and I, that is all."

This was all unknown to her. She had loved him, the boy with the extraordinary eyes, the boy who played so magnificently, who laughed so much. But now there was roused something more than these. The piano player was gone, he did not laugh, his eyes had never quite glowed like that, there was in his face something she had never seen yet. The woman had awakened the man; this was his first full moment of consciousness. And, like all women for the first time face to face with the lover and the beloved, she was afraid. She had not till now seen his full fire.

"I am frightened!" she cried. "What have we done?"

But his answer came back like an echo to what she had not said, but what was behind her words.

"Frightened?" he said. "Oh, Stella, not of me, not of the real me?"

She gave a little laugh, still mysteriously nervous.

"You were a stranger," she said. "I never saw you before."

Martin gave a great happy sigh, and the authentic fire leaped to and fro between their eyes. "You are quite right," he said. "I was never this before. But you are not frightened now?"

This time her eyes did not waver from his.

"No, Martin," she said.

But there was no more privacy possible here. Stella had been quite right; there was a party going on, and at the moment a great burst of applause signified the end of Karl Rusoff's performance. Stella started.

"There; I told you so," she said. "Now take me to my mother, she will be waiting for me."

Martin frowned.

"Cannot she wait?" he asked. "I too have never seen the real you before."

"No, dear, we must go. There is to-morrow, all the to-morrows."

"And to think that it has only been yesterday until this evening," he said. "There is Lady Monica, looking for you."

Lady Monica had a practised eye. She kept everything she had in excellent practice; there was nothing rusty about her.

"Stella, dear, I've been looking for you," she said. "Are you better? Has Mr. Challoner been taking care of you?"

That was sufficient.

"Stella says I may," said he.

Lady Monica checked her exclamation of "Thank God," as being a shade too businesslike.

"Ah, dear Mr. Martin," she said. "How nice, how very, very nice! Stella, my dearest! How secret you have been! Come, darling, we must go. I can't talk to either of you in this crowd. But how nice! We shall see you to-morrow? Come to lunch, quite, quite quietly."

Stella looked at him.

"Yes, do, Martin," she said. "I will take you back after our skate."

"Ah, I had forgotten," he said.

She laughed divinely.

"But I had not. And you will be kind to me, as I asked you?" she added.

He dwelt on his answer.

"I kind—to you?" he said.

CHAPTER XII.

It was a March day of glorious windy brightness, a day that atones and amends with prodigal open-handed generosity for all the fogs and chilly darknesses of autumn and winter. Heavy rain had fallen during the night before, cold, chilly rain, but an hour before morning it had ceased, and a great warm boisterous wind came humming up from the south-west. Like some celestial house-cleaner it swept the clouds from the face of the sky, and an hour of ivory starlight and setting moon ushered in the day.

That same wind had awakened Helen with the sound of the tapping, struggling blind drawn over her open window, and with eyes suffused with sleep she had got out of bed to quiet the rattling calico by the simple process of rolling it up. And, having rolled it up, she stood for a moment at the window, her hair stirred by the wind, drinking in the soft cool breath of the huge night that blew her night-dress close to her skin. The clean smell of rain was in the air, but the sky was all clear, and to the east behind the tower of Chartries church, the nameless dove-coloured hue of coming dawn was beginning to make dim the stars. Then she went back to bed with a vague but certain sense that some change had come—winter was over: in her very bones she felt that.

Gloriously did the morning fulfil her expectations.

White fleecy clouds, high in the heavens, bowled along the blue, their shadows racing beneath them across the brown grass of the downs; the wind, warm, and pregnant with Spring, drove boisterously out of the west, and the sun flooded all that lived in a bath of light. Round the elms in the churchyard there had been wrought that yearly miracle, that mist of green leaf hovering round the trees, and paler and more delicate it hung round the slim purple-twigged birches in the woods that climbed up the hillside beyond to Chartries. Here after breakfast her path lay, for she had a parish errand to an outlying hamlet beyond, and with eye and ear and nostril and open mouth she breathed and was bathed in the revivification of Spring. That morning, so it seemed to her, all the birds in the world sang together, thrushes bubbled with the noise of chuckling water and delicious repeated phrases of melody, as if to show, brave musicians, that the "first fine careless rapture" is perfectly easy to recapture, if you happen to know the way of the thing: blackbirds with liquid throat and tawny bill scudded through the bushes, above, swifts chided in swooping companies, and finches and sparrows poured out staccato notes. One bird alone was silent, for the nightingale waited till summer should come and love.

That filled the ear; for the eye there were blue distances, blue shadows of racing clouds, the sun, and, more near, the budding trees, and in the dingle below the woods of Chartries a million daffodils. Helen had forgotten that they were there, waiting for her, and she came on them suddenly, and stood quite still a moment with a long pause of pure and complete delight. The place was carpeted with them; they all danced and shone and sang together like the morning stars. And as she looked her eyes grew dim with happy tears.

"Dear God," she said, "thank you so much!"

Yes, indeed it was Spring, and as she walked on she repeated the word over and over again to herself, finding a magic in it. It was everywhere; the sky and the sun were full of it, it burst in those myriad blossoms from the dark wholesome soil; it was Spring that set this good wind blowing, it bubbled and chuckled in the chalk stream with its waving weeds and bright glimmering beds of pebbles. Above all, it was in her own heart on this glorious morning, till she thought it must almost burst too, so overflowing was it with sheer unreasoning happiness.

Indeed, Martin had been quite right when he had told Karl how happy she was, and though she did not reason to herself about it the cause was abundantly clear. For the last six months she had lived at home, through days and weeks of ever-recurring difficulties, and with each, as it presented itself, she had dealt smilingly, patiently. She had made up her mind on her visit to Cambridge that her duty was clear and obvious: nothing striking or picturesque was in the least required; she was neither going to renounce her future happiness, nor, on the other hand, to throw all else aside and grasp at it. No heroic knot-cutting measures of any kind were indicated, except the quiet, unobtrusive heroism of taking up again, quite simply, quietly and naturally, all the straightforward, familiar little duties of her home life, which again and again she had found so tedious. Nor had they been in themselves less tedious, only—here was the difference—she had ceased to look upon them from any point of view except one—namely, that it was quite distinctly her business to do them. That she had found to be sufficient: it was enough day by day to get through with them, without expenditure of thought as to whether they were distasteful or not; and her work, her daily bread, had somehow been sweet and wholesome and nourishing. Truly, if, as Karl had said, Martin had been growing out of knowledge, his twin also would be scarcely recognisable.

And bread, bread of the soul, had come to her, her table had been laid in the wilderness, and happiness, royal inward happiness of a very fine and unselfish sort, in the midst of a thousand things which made for unhappiness, had blossomed in her. A thousand times she had been tempted to say, "It is doing no good: why should I put off what is waiting for me, when my renunciation does not help father in any way?" But a thousand times she had just not said it; and now at the end of these difficult months, she could, without egoism, look back and see what infinite good had been done. That her father should in any way alter his own convictions about her marriage she had never expected; but what had been gained was that he saw now, and consciously saw, that she was, in the very simplest language, "being good."

But it had been difficult enough for all concerned, except perhaps for Aunt Clara, who was scarcely capable of emotion, and often Helen's heart had bled for her father. It had been most terrible of all when Martin had joined the Roman Church. His letter to his father—Helen winced when she thought of it now—had arrived on Sunday morning, and he had found it on the breakfast table when he had come back from the early celebration. It was a manly, straightforward letter enough, stating that he had not yet gone over, but had practically determined to. If his father wished he would come down to Chartries, and talk it over with him, and give to his advice and counsel the very fullest possible consideration. And at the end he expressed very bluntly and sincerely, as was his wont, the sorrow and the pain that he knew the news would cause his father.

The sheet fell from his hand, and Helen, who was making tea, looked up. She saw the colour rise in her father's face; the arteries in his neck and temples swelled into cords, and his eyes, with pupils contracted to pinpricks, looked for the moment like the eyes of a madman. Then he spoke, his voice vibrating with suppressed furious anger.

"Martin is going to join the Roman Church," he said. "From the day he does so, Helen, never speak to me of him again. He is dead to me, remember."

That was a week before Christmas, and for more than a month after that Martin's name had literally hardly crossed his father's lips. The boy had come down to stay with his uncle once, but Mr. Challoner had absolutely refused to see him. He had even wished Helen not to; but on this occasion, for the only time during all that long winter, she had quietly but quite firmly disobeyed him. It was then first, too, as one looking down from barren rocks of a mountain range, that she saw, though still far off, the harvest that was ripening in these long patient months of her living here with her father. Before going to Chartries she had thought best to go into his study and tell him that she was doing so.

"I am going to see Martin," she said, wondering and very nervous as to how her father would take it. "And I wanted to tell you, father, before I went, that I was going."

Mr. Challoner was writing his sermon, but on her words his pen paused; then he looked up at her.

"Very well, dear," he said; "you know my feeling about it, but it is a thing in which you must do as you think right. And, Helen—" again he paused, and his eyes wandered away from her and were bent on his paper, "tell me when you come back how the lad looks: if he seems well."

She came closer to him; this was the first sign he had shown that he recognised Martin's existence.

"Ah, father, come with me," she said.

But he shook his head.

"No, dear; no, dear," he said, and went on with his work.

But on this March morning of windy brightness, what gave the *comble* to her happiness was the talk—the first intimate one for all these weeks—which she had had with her father the night before. She had gone to her room as usual after prayers, but finding there some parish work, concerning outdoor relief, which she ought to have done and taken to him the day before, she sat up for nearly a couple of hours, until she had finished it. Then, with the papers in her hand she went down to his study.

"I am so sorry, father," she said. "You told me you wanted these yesterday, and I absolutely forgot to do them. They are finished now."

He looked up in surprise.

"Why, Helen," he said, "it is after twelve. You ought

to have been in bed long ago. Have you been sitting up to do them?"

She smiled at him.

"Why, yes," she said.

He took them from her.

"You have been a very good daughter to me, dear," he said.

He paused, but Helen said nothing, for his tone showed an unfinished sentence. And the pause was long; it was not at all easy for him to say what followed.

"And I have been often and often very difficult and very hard all these months," he said. "But will you do your best to forget that? Will you try to forgive me?"

She went close to him, very much moved, and laid her hand on his shoulder.

"Ah, don't cut me to the heart," she said.

"But promise me, if you can," he said.

Yes, it was true: he had often been difficult and hard. And she answered him.

"Yes, dear father," she said. "I promise you that with my whole heart. And in turn, when May comes, will you try not to think too hardly of me? I have tried to be good."

She sat down by his side, looking rather wistfully at him.

"I have been wanting to talk to you often before about that," he said, "so let me say once and for all what is in my mind. I disagree with you, as you know, vitally, essentially, and I believe that God tells me to disagree. But now I believe also, dear,—and this your goodness and your sweet patience all these months have taught me,—

that God tells you to do as you are going to. How that is I do not understand. Perhaps that doesn't matter so much as I used to think. But He fulfils Himself in many ways. And there too I have very often thought that He had to fulfil Himself in my way. It is you who have made me see that, I think."

Helen raised shining eyes to his.

"You have made me very happy," she said.

"And what have you done for me? There were certain days, dear, during this winter which I do not see how I could have got through without you."

Here was an opportunity for which Helen had often sought.

"Martin?" she asked. "Oh, father, I wonder if you want Martin as much as I do."

The strength and the tenderness died out of his face, leaving it both helpless and hard.

"I can't see him," he said quickly. "I dare not. Some day perhaps, but if I saw him now I should say, I could not I know help saying, what I feel. If that would do any good, I would say it, but it would do none. I should only—I should only frighten him," he said, with an accent infinitely pathetic.

She left him then without more words, for all this winter she had been learning every day and all day long the divine and human gospel of patience in dealing with people, the patience that teaches us not to pull buds open, however desirable it may be that the flower should unfold; that is content to do its best with them, and wait for results without the desire even that they should come quickly. Till this evening, as has been said, Martin's name had scarcely been mentioned by his father; and it was something, after this bitterness of long silence, that he should be able to say, "not yet, not now." Pity also, pity with hands of healing, had entered at last into that stern, upright, God-fearing soul, filtering its way like water through dry and stony soil, a very exiguous trickle it might be, but cool, liquid, refreshing. How hardly it had won its way there Helen but guessed dimly, he alone knew. For day had succeeded day, and week, week, and all day and all week he had wrestled blindly, hopelessly, with the misery that Martin had brought on him, unable for all his efforts to find any possible justifying cause for what he had done, which seemed to him as wanton and as wicked as violent crime. To his Puritan mind, Martin's reason, namely, the craving for and the necessity of beauty and poetry in religion was as unintelligible as a page in an unknown language; not knowing at all what that craving meant, any more than he knew what homicidal mania meant to a maniac, he could not in any degree whatever feel or appreciate its force. And for the sake of this his son had left the mother Church, and embraced the heresies, the abominations, the idolatries of Rome. Such was his sober literal view: the Roman Church was idolatrous, and for idolaters was the doom appointed, revealed by God, believed by him. And there stood Martin.

For weeks nothing had come to sweeten the bitterness of these dark waters: his suffering was as unintelligible to him as is pain to a dumb animal: he could not guess what it could possibly mean. That fierce anguish, like a flame, had burned up for a time in its withering breath all human affection: he had hated Martin for what he had done. Shocking as that was, he knew it to be true, and his hate seemed somehow justified. There were things, there were actions and passions which he was bound to hate: and so filled was he with this conviction, that human affection, human love could find at first no place in his mind: it was turned out, evicted. But now, like a dog beaten and driven from the house, it was beginning, so Helen thought, to creep noiselessly, stealthily homewards again. So she was content: she did not even want to hurry it.

And this morning Spring was here, too, and the daffodils danced.

From the dance of daffodils the slope rose steeply upwards through the hanging woods of Chartries, and her path lay by the bushes in which last summer Frank had found the trapped hare. Here, as always, she went slowly, telling over in her mind, like the beads of a rosary, the history of those hours. Then, raising her eyes, she saw him, Frank, standing a little way up the path, looking at her.

Involuntarily her heart leaped to him, and holding out both hands she quickened her step as if running to him. That first movement she could no more help than she could help the fresh blood springing to her cheeks. But at once almost she recollected herself and paused.

"Ah, Frank," she cried, "you shouldn't have come here. You know you shouldn't!"

He came no nearer.

"No, my darling," he said, "but I couldn't help it. It is not your fault, you have not broken your promise. I only had to see you, just see you. I think it was the Spring that made me do it. I am with your uncle for just one night. See, there is this for you from Martin."

He held out a note for her, standing a little aside, so that the path was clear for her to pass on her way. But as their fingers met, she lingered and hung on her step, still not looking at him. She tried, she tried her best to pass on, but she could not; her eyelids swept upwards, and she looked at him. Then which of them moved first neither knew, but next moment his arms were round her, and he kissed her. And, alas, her struggle to get free was very faint: her tongue protested, but not very earnestly.

"Ah, let me go, let me go," it said.

"I can't. Helen, it was here that——"

"I know," she said. "I come here every day; I knew I should meet you here some day. And this of all days, the first of Spring. Oh, Frank, let me go. I love you: is not that enough? And it is not for long now."

"No, my darling, it is not long now."

"And—and, it has been so long. And I have wanted you so much."

She disengaged herself quietly, but in a way that made it impossible for him to hold her, from his arms.

"Good-bye," she said. "You ought never to have come. And—oh, my darling—I thank you so for coming."

For one infinitesimal moment she looked at him again, then with her quick light step she went on up the path with Martin's letter in her hand, and never looked back. She did not pause till she reached the top of the wood, but as she walked she listened for and longed for and yet dreaded to hear footsteps behind. But none came, she had made her meaning too clear for that (and how she wished she had been less explicit!), and having arrived at the top, she slackened her pace, and opened Martin's letter. It was very short, a couple of lines only, announcing his engagement to Stella, and asking her to tell his father. And with that Spring was complete.

Upwards again lay her path, but no more among trees and sheltered places, but high over the broad swell of the short-turfed downs, where shadows of clouds ran glorious races. Something in the huge view and the large sky chimed in wonderful harmony to the girl's mood; all was so big, so untainted, so full of light. Beneath her foot the dead autumn turf still stretched in brown tufts and patches, but springing up in between were the myriad shoots of the young grass, and even since yesterday, she thought, the tone of the colour was changed. Till to-day all had been grey and brown, all still pointed backwards, winter-wards, but this morning it was different, and the million sprouting lives shouted, "Look forward, look forward! For, lo, the winter is past and the time of the singing bird has come." "Ah! song of songs," she thought, "indeed it is so."

Martin! There were no words into which she could put what she felt, any more than the pervading sunlight could be put into words. It was there, a great, huge, exultant presence that flooded everything. Ah! the beloved twin! Why, it was only a few years ago that he was in Eton jacket and broad white collar, and sang treble. And she? Well, yes, she was in short frocks about the same time. Yet had not she, half an hour ago, down in the wood below her, where the young leaves hung like a green mist round the purple branches of the birch, felt a loving arm round her, and kisses on her face? Oh, it was very wrong of Frank. No, not wrong of him—he would have stood aside; he did stand aside to let her pass. It was very wrong of her. But at that moment she could not pass by—it was as if her power of movement had been paralysed. Yet she was not in the least degree ashamed of herself, and she looked forward with a certain secret glee to telling her father—for that had to be done—for so by speaking of it she would live it over again. "No, that was not all," she said to herself, rehearsing question and answer. "He kissed me."

Sunlight, and larks invisible, and the shadows of clouds that coursed over the downs. And some distance off a tall figure moving towards her rapidly, a figure she easily recognised. They came nearer and met; her hat was in her hand, her hair tossed over her forehead, and there was Spring and the sure promise of summer in her face. And in her father's, too, there was something of that infused joy; his hand held a little bunch of primroses which he had plucked as he walked.

They met without words, but with smiles, the unconscious smile that the morning had made.

"Well, Helen!" said he.

He came close to her, and with his neat precision put the primroses into her hat.

"You ought to pin them," he said. "They will fall out."

She laughed.

"Ah! nothing can fall out to-day," she said. "Don't

you feel it, father? Spring, Spring—and—oh, the daffodils! And I have news."

Then her face sobered suddenly.

"Two pieces of news," she said, smiling again, unable not to be gay. "The first is of Martin; he is engaged to be married. He asked me to tell you. Stella Plympton, whom you met here. He wrote me just a line, asking me to tell you."

Her instinct was right to repeat that. Sharp as a knife, a father's jealousy had pierced him. He should have been told first; whatever his disagreements with Martin, he, his father, ought to have been told first. But that passed in a moment.

"Martin?" he said gently. "The boy?"

"Yes, I thought of it like that. But he is really—oh, ever so old. As old as I am."

Mr. Challoner's face relaxed.

"I had forgotten," he said, "an immense age! What next, Helen?"

She looked up at him.

"Is that all you have to say?" she asked, feeling suddenly chilly and disappointed.

"You think I am hard, Helen," he said. "I try not to be. But what next?"

Yes, it was chilly on these upland downs. She put her hat on.

"Just this," she said. "I met Frank half an hour ago. He gave me Martin's note. I did not expect to see him. As far as I am concerned it was quite accidental. I had no idea he was here. I had promised you not to see him. That I could not help."

She stopped: drew a long breath and went on.

"I suppose I could have helped the rest," she said. "I suppose it was that I did not choose to help it. He stood aside for me to pass. But—but I did not pass. I went to him. I let him kiss me. He stood there with me. I thought I could not help it. Indeed, I thought that."

For a moment Mr. Challoner's hardness, his involuntary condemnation of weakness of any sort, of failure to keep a promise, returned to him, mixed with a very ugly thing, suspicion.

"And this is the first time you have seen or spoken to him, or had any communication with him?" he asked.

Helen raised her eyes to him in quiet surprise. No trace of resentment or sense of injustice was in her voice.

"Yes, of course," she said. "I should have told you otherwise."

He looked at the sweet patient face, struggling for a moment with this worse self of his, which yet was so upright, so devoted.

"I know you would," he said at last. "I don't know why I asked you that."

Helen laughed.

"Nor do I," she said.

"You and he have been very patient, Helen," he said.

"Yes, till this morning I think we have," she said. "But to-day, perhaps the Spring was too strong for us both. Is it not in your blood this morning, father? It is in mine."

He smiled at her gravely.

"And a very suitable thing," he said. "And summer comes next for you. For you and Martin."

"Yes, Martin too," she said, with an appeal in her eyes. "Oh, father, can't we all be happy together again? We used to be."

Mr. Challoner stood silent a moment, a sort of aching longing for all he had always missed in Martin, and a dim, bitter regret for all his own missed opportunities of making the most of the human relation between himself and his son, rising suddenly within him. And he spoke with a terrible quiet sincerity.

"I don't think Martin used ever to be happy with me," he said. "Once he told me he was not happy at home. I don't think that he ever was. It was perhaps the fault of both of us, but it was certainly mine. I should have done somehow differently. I think we never understood each other. Nor can I understand him now. It is sad. I cannot reconcile what he has done——"

He broke off again.

"There, dear, you must be getting on your way," he said, "and I must be getting home."

But she detained him a moment more.

"Won't you give me a little hope?" she said. "I thought last night that perhaps, perhaps soon— And this news this morning—"

But her father disengaged her hand.

"I shall of course write to him," he said, "and congratulate him. She is a very charming girl. I think Martin is most fortunate."

"Martin is very charming too, remember," said she.

Mr. Challoner walked swiftly homewards after Helen had left him, feeling strangely and deeply moved by the news. He felt somehow that his children were his children no longer; all the responsibility for them had passed into other hands, and they themselves light-heartedly, eagerly, were now taking on themselves the responsibility for others. He had thought of them always as a boy and a girl, each bound to obedience to his will, dependent on him, without any real individual existence of their own. But within the last year first one and now the other was passing out of his reach. Helen first and then Martin had acted for themselves in direct defiance, not only of his wish, but of that which was the mainspring and motive of his life. She, it is true, but these months of quiet, normal life at home, had made a great change in him; her disobedience to him personally had vanished from his mind, and as he had told her last night, though he believed no less strongly than before that his conviction with regard to her marriage was the will of God for him, he believed also, though he could not understand how, that she, too, was acting consonantly to that same will. But with regard to Martin, however he looked at his conduct, or whatever possible interpretation he tried to put on it, he could not see light. Martin was trivial, superficial, not in earnest about religious matters, just as he had been in the rest of his education. Nothing, except music, which Mr. Challoner could not frankly bring himself to regard as anything but a mere æsthetic fringe, a mere ornament of life, had ever touched him deeply. He had no depth, no seriousness. And now that boy, that child, was going to be married, to take upon himself with the same light-hearted *insouciance* all the responsibilities of a husband and a father.

How strange that they were twins! Helen developing

every day in patience, dutifulness, love, and Martin still thoughtless, bent only on the personal gratification of his musical tastes, and willing, so Mr. Challoner bitterly put it to himself, to leave the English Church, the mother of his faith, for the sake of a hymn tune! He would write to him, as he had said, but even now he could not see him. For he knew himself well, and recognised, though he scarcely wished to cure, his own impatience, his anger at one who seemed to him to be going wrong wilfully. On a point like this he could make no concession, for any concession implied a failure of loyalty on his own part to his creed.

He had by this time entered the woods round Chartries where the path was wet and a little slippery under the trees, causing him to abate the briskness of his pace. How different, how utterly different Helen had proved herself! If only she could see the question of her marriage as he saw it, how would his whole heart rise up in thankfulness! For though he admitted here that both he and she might be right, he was still full of disquietude and anxiety about it. Then, suddenly turning a corner, he found himself face to face with her lover.

For a moment neither spoke; to Frank it seemed that if words even of commonplace greeting were to pass between them, it must be for Mr. Challoner to make the beginning, while to the elder man the sudden shock of seeing him, inevitably awakened again for the moment the horror and bitterness of their last interview. Under that his mouth was compressed and tightened, a gleam almost of elemental enmity shone in his eyes, and it seemed to himself that he would pass by Frank with averted head. But then over that, veiling and softening it, there rose all that he had been learning this winter, all that Helen had been teaching him, and as he came close to Frank he paused. Then, with an effort that cost the proud man something, he put his lesson into practice, and held out his hand. And the strength and the big lovableness of the man was offered with it whole-heartedly.

"We shook hands last time we met, Lord Yorkshire," he said. "Will you not let me shake hands with you again?"

That done, that effort made, the rest was easier, for all that was generous and sympathetic in Frank responded.

"Thank you," he said simply. "And I am not exaggerating, Mr. Challoner, when I tell you that I know nothing in the world that could have happened to me which could give me so much pleasure as this."

Mr. Challoner still retained his hand.

"Do you know, you are a very good fellow?" he said. "You are very generous to me. So has Helen been. I cannot tell you what she has been to me all this winter. And I thank you very much for letting her be with me, for not urging her otherwise. You have made it all as easy for her as you could. You have been very unselfish, both of you. And I have been making it very difficult for her."

Frank was a good deal moved. There was a very noble and a pathetic sincerity about this.

"I think you wrong yourself," he said. "I am sure you wrong yourself. We have all tried to—well, to do our best. And we all three of us know that."

But Mr. Challoner had more to say.

"I ask your forgiveness," he said simply, and his voice trembled a little.

"Ah, don't do that," said Frank.

They stood there together a moment longer, under the flecked sunlight filtering through the trees, suddenly brought close again, just as they had been in that dreadful hour when Helen's weakness made them forget all else. But now the reconciliation went far deeper than it had gone before. Then they had joined hands in ministering to the physical suffering of one they both loved; but now they joined hands over an appreciation, not of weakness, but of strength. The bond between them was no longer a thing that could easily break; poles apart as they still were, that golden thread could scarcely be snapped.

"I met Helen just now," said Mr. Challoner at length. "She told me she had seen you, dear girl. She told me also the news from my son. Are you busy? Will you walk with me a little way?"

Frank turned at once, and they went on down the steep path towards the Rectory.

"Have you seen Martin lately?" asked his father.

"Yes, I see him constantly in London."

"Then can you tell me about him? *What* is he? That is the thing I puzzle and pray over. He joined the Roman Church, as you know, at Christmas. I don't think anything ever pained me more. But I should be very glad to know that he is in earnest about it. Or does he take it as he takes everything else? Do you understand it?"

"Yes, I think I do," said Frank, and paused a moment. "It is this. Martin demands beauty in all that is real to him. That is the ruling instinct in his nature. And in matters of religion, the Roman Church seemed to him to supply that more than the Church he left."

"And it was for that he threw it over?" said Mr. Challoner. "And without regret or struggle even?"

"He regretted very sincerely the pain it would give you," said Frank.

Mr. Challoner waved this aside.

"That does not matter," he said. "But otherwise without a regret?"

Frank let his silence unmistakably answer that before he went on.

"I know you will excuse me," he said; "but I don't think you quite realise what Martin is, or how the artistic instincts dominate him. Till he fell in love I don't think he ever had any very poignant emotion apart from them."

Mr. Challoner's face got even more grave.

"Simply then," he said, "he puts them above the love of God. I do not understand how a Christian can do that. And I do not want to understand it," he added.

They had reached the Rectory, and Mr. Challoner paused on the terrace.

"Is he a good boy?" he asked suddenly.

"Morally? Yes, I am sure of it."

"How do you know that?" asked his father.

"Because I know his opinion about immorality. He feels very strongly that it must blunt the artistic sense."

Mr. Challoner winced as if in sudden pain.

"Ah," he cried, "is that all? Dear God, is that all?"

"The result in the way of conduct is identical," said Frank quietly.

"Yes, yes; but are we not taught that works without faith are dead? Ah, I beg your pardon, indeed I do, my dear fellow. I spoke without thinking. I was thinking only of my poor Martin. Pray forgive me. And is he happy, do you think?"

"Yes, quite extraordinarily happy. He has fallen in love too with the same white ardour that he brings to everything which appeals to him."

Mr. Challoner considered this a moment, and then faced Frank.

"I want your opinion, Lord Yorkshire," he said. "Do you think that any good purpose would be served by my seeing Martin? I ask you for your candid opinion, whatever it is or implies."

"I think it depends entirely on yourself," said Frank.

"You mean—ah, pray tell me quite straight out. I shall be very grateful."

Frank looked at him with very real pity. What he was going to say seemed very cruel, but it seemed true.

"I mean this, Mr. Challoner," he said, "that if you are quite certain that the sight of Martin, or the possible issues into which talk may lead you, will not again embitter you against him, you had far better see him. Why not? There is all to be gained. But if your reconciliation cannot be complete, if there is a chance of your getting angry with him, and—frightening him, you had better not. You asked me to tell you straight."

"You think he is afraid of me? Has he told you?"

"I cannot help knowing it. If he has told me, you must take my word for it that he has not told me in any disloyal way. And if I have hurt you, I am very sorry."

"No, I thank you for telling me," said Mr. Challoner. "I think you are right. I am afraid it is better I should not see him yet."

He smiled rather sadly.

"I am afraid I have a great deal to learn yet," he said. "I must take myself in hand. But I dream about him, Lord Yorkshire, so often. And always almost in my dreams I say things to him that frighten him. Sometimes it is true we are great friends: those are beautiful nights, and I thank God for them. I so long to see his dear face again."

"Those beautiful nights must find fulfilment in many beautiful days," said Frank.

"Yes, I hope that. It is still possible. He was such a bright little fellow when he was small. Always quick, always laughing. I had many plans for him. I think all my life I have been rather too ready to push other people into places I think suitable."

They had come to the far end of the terrace again, when from inside the Vicarage the gong sounded for lunch. Frank's back was towards the house, but the Vicar looking up, saw Helen, still hatless, coming towards them across the lawn. And all the happiness of the morning, when she saw these two together, all the spirit of Spring, quivered and concentrated itself into one rose-coloured point of joy. That was the best moment to her in all the days of Spring that were yet to come.

"You will stay to lunch, Lord Yorkshire?" said he.

"Thanks so much, but I am afraid I ought to get back to Chartries: I said I would be back."

Mr. Challoner waited till Helen was close to them.

"Perhaps if Helen adds her voice to mine," he said.

He turned quickly and saw her. And there was no need of words, but once more the three stood together, hands clasped. This time the Vicar did not go back alone to his empty room.

CHAPTER XIII.

MARTIN was seated alone with Stella in the drawing-room of her mother's house, eating muffins thoughtfully, but rather rapidly, while she poured out tea.

"Fancy," he said, "it is only a week ago since—since the party at Lady Sunningdale's, since I knew."

"Knew what?" asked Stella quite unnecessarily.

"Ah! I only know one thing now. I think I have forgotten everything else."

"Say it then," said she.

"That I love you? Are you not tired of hearing me say that yet?"

She smiled, brought him his tea, and sat on the arm of his chair.

"I can't believe that a woman can ever be tired of hearing that, if the right man says it. Oh, Martin, how lucky it was you, and that it was I!"

Martin put his tea-cup down, having drunk with amazing speed.

"Why, who else could it have been?" he said. "How could it have been otherwise?"

"No, I suppose not. Yet you didn't know, as you call it, for a long time. Supposing you had gone on not knowing?"

He leaned back in his chair looking at her, his black eyes shining in the firelight.

"And when I did know, I frightened you," he said.

"Yes, a little. But I loved it. You see I had never seen you really in earnest before, except when you were playing. You always put everything you had or were into that."

"I know. That is what Karl Rusoff told me. He told me to experience all I could, because it would all go to make me play. He calls it spiritual alchemy, like when you put a plant in the earth and water it, the earth and the water are somehow turned into the blossom of that plant, while another plant would turn them into a different flower. In fact, darling, you are going to come out of the ends of my fingers, whereas, if I were a great Greek scholar you would become iambics."

He looked at her and his smile deepened into gravity.

"Oh, Stella, Stella," he said, "did the world ever hold anything like you?"

She leaned back till her face was close to his, and put her arm round his neck.

"Yes, yes, do that with me!" she said; "absorb me, let me become part of you. Indeed, I want no other existence at all. Do you know the Persian legend, how the lover knocked at the door of his beloved, and the beloved said, 'Who is that?' and he replied 'It is I.' And the one inside said, 'There is not room for two.' Then he went away again, and came back after a year and knocked again. And again from inside the voice said 'Who is that?' But this time he said, 'It is thou.' So the door was opened and he went in."

"That is beautiful," said Martin. "But, my word, fancy being able to become music. And suppose one happened to become a song by Gounod. Only that isn't music," he added.

Stella felt somehow suddenly chilled.

"Promise me I shan't become a song by Gounod," she said.

Martin looked at her in silence a moment. She had risen rather abruptly from her position, and was again sitting upright on the arm of the chair.

"And what do I become?" he asked. "What do you make of me? It is thou, remember."

Something that for the last three days had hung mist-like in Stella's mind suddenly congealed, crystallised, became definite.

"I don't want you to become anything," she said. "But I want you to be—I want you to be entirely yourself. I want you to get below your own surface, to dive into yourself, to find pearls. And then to let me wear them."

"You mean I am shallow?"

"No, dear, I mean nothing of the kind. But, oh, Martin, don't misunderstand me. All you have got from life, all you have gained, all you are, you treat as fuel—you have said it—to burn in the furnace of your one passion—music."

Martin admitted this with a reservation.

"That was true," he said, "till just a week ago."

Stella rose from her place; sitting close to him like that she could not say what she meant to say. Personal magnetism, her love for that beautiful face, prevented her. So she went to the hearthrug, under pretence of poking the fire, and stood there with her back to it, facing him. Then she spoke more quickly, with a certain vibration in her voice.

"And this last week," she said "a new and wonderful piece of music was discovered by you. Yes, I put myself as high as that. But am I more than that? Am I really?"

Martin's forehead wrinkled slightly. Had it not been Stella who asked him this, he would have said the question was unreasonable. But before he could reply she went on.

"Ah, dearest," she said. "I asked you just now to absorb me, to make me you. But I will not flow out of your finger-tips. Oh, I know you only said that in jest, but in jest sometimes one strikes very near to truth. Have you thought what you are to me, and what, if I am anything, I must be to you? Something absolutely indispensable, your life, no less. Now, supposing chords and harmonies were dumb to you for ever, what would be left of you? Tell me that."

Martin's expression grew more puzzled. It was as if she asked him some preposterous riddle without answer. How could he compare the two?

"How can I tell?" he said. "I suppose I should somehow and sometime adjust myself to it, though I haven't the slightest idea how. I can't imagine life, consciousness without them!"

"And if I went out of your life?" she asked unwisely, but longing for some convincing answer.

In reply Martin got up, and went close to her.

"You have often called me a fool," he said, "and you have often called me a child. I am both when you ask me things like that. But this foolish child speaks to you, so listen. He does not know what it all means, but he loves you. He knows no other word except that. Is that not enough? If not, what is?"

Then once again the mastery of man overcame her. She wanted him so much, more than any answer to her questions. The subtleties into which she had tried to draw him he brushed aside; her woman's brain, her woman's desire to hear him say that she was all, had spun them deftly enough, but he blundered through them somehow, like a bumble bee through a spider's web, and came booming out on the other side. Theoretically, anyhow, if he had been a woman, they must have caught him, he must have struggled with them, felt their entanglement. As it was she had failed. Probably he labelled her fine spinnings "silly" in his own mind. But he proceeded through them—still frowning a little.

"You ask me impossible riddles," he said. "You might as well ask me whether you would sooner tie your mother to the stake and burn her, or me. My darling, there is no sense in such things. Surely one can be simple about love, just because it is so big. I know I love you, that is enough for me. I told you that I know nothing else. That is sober truth. But I cannot weigh things in balances. And what is more, I won't. Now kiss me: no, properly."

It must therefore be inferred that he got his way in this matter, for when, two minutes later, Lady Sunningdale made her untimely appearance, the two were again seated, Stella this time in the chair, and Martin on the arm.

"But famishing," she said. "Yes, tea please, dear Stella. Martin, you monster, I haven't seen you for days. Why I haven't taken to drink I don't know, owing to all the dreadful things that have been happening. Would you believe it, Sahara had two puppies, but she couldn't bear them, so she ate one and starved the other. Well, it's all over, but nobody in the house has had a wink of sleep for the last week. And so you are going to give a concert at last, Martin! I shan't come. I hate my private property being made public."

"But charity," said Martin.

"My dear, I know perfectly well what charity and St. James's Hall means. It means guinea tickets. Charity should begin at home, not at St. James's Hall. However, I daresay you will appropriate all the proceeds. So near the Circus too. Really, Piccadilly Circus is too fascinating. I should like to have a house in the very centre of it with a glass gallery all round, and really see life. Yes, one more piece of muffin, not for myself, but for Suez Canal. Suez Canal is so lonely, poor darling, without Sahara, but there is muffin *quand même*. Naughty! I'm sure the servants feed him. And so everybody is to be married in May! Fancy the Bear coming round like that—even Bears will turn—about Helen and Frank. Apparently they are quite inseparable, the Bear and Frank I mean, and tie each other's bootlaces, and are converting each other to Christianity and Atheism respectively. Bears and buns! Frank is a bun, and the Bear has decided it is worth climbing up a pole to get him. I think it is a mistake to have said that. Besides, it is absolutely untrue. The Bear wouldn't climb a yard to marry Helen to the Czar. How terrible Russia must be with everything ending in 'owsky'! I tried to flirt with the Bear myself, and had no success of any kind whatever. Dear Suez! No Sahara. The world is a desert without Sahara. But mayn't I tempt you with a small piece of bun with sugar on the top? How depressing marriages are!"

Lady Sunningdale sighed heavily.

"What is the matter?" asked Stella sympathetically.

"I don't know. Dearest, that Louis XVI. clock is too

beautiful. I wish I were a millionaire. Yes. I think I am depressed because everything is going exactly as I planned it. There is nothing so tiresome as success. You two children sitting there, Frank and Helen, all my own ideas, and all going precisely as I wished. You are my idea too, Martin, a figment of my brain. I invented you. And you are going precisely as I wished. Everyone says nobody ever played the least like you. But the Bear is still in a rage with you, is he not? That is so English. English people are always in a rage about something, the state of the weather, or France, or their children. I never get in a rage. I have no time for that sort of thing. Stella, dearest, I think it will have to be you to go down to Chartries next, and induce the Bear to be propitiated. Heavens, how dreadful it must be to have a very strong sense of duty! It must be like toast crumbs in your bed, after you have breakfasted there, when one can't lie comfortable for five minutes together."

"No, I am the next," said Martin. "I shall be staying with my uncle at Easter, and shall try to see my father then. I dare say it will do no good."

"Do you really care?" asked Lady Sunningdale. "I really don't see why you should. He is unreasonable; I shouldn't worry."

Stella turned to Martin with a certain air of expectancy.

"Yes, I do care," he said; "I care horribly. I care every day. I hate being on bad terms with anyone. I hate anger and resentment. Besides, he is my father. That is why I care so much," he added with a little quiet air of dignity, for he had not wholly liked Lady Sunningdale's remarks.

"That was one of Nature's most extraordinary conjuring tricks," she said. "People talk of heredity, but put all the fathers of England in a row, and ask anyone to pick out Martin's. The better they know either of you, by so much the more will they pick out Mr. Challoner last of all!"

Martin got up.

"Ah, don't let's talk about it," he said; "it is not agreeable. I wish I could laugh about it like you, but I can't."

Then with a quick intuition he turned to Stella.

"One can't do any good by talking about it, can one?" he asked.

Something still jarred on the girl, due partly to their talk before Lady Sunningdale came in.

"You have admirable commonsense," she said.

Lady Sunningdale caught on to this with her usual quickness. She knew for certain from Stella's tone that something had gone just a shade wrong between them.

"And you find it rather trying, do you not, dearest Stella?" she said. "Of course, Martin is the most trying person in the world, and if it wasn't for his ten fingers he would be absolutely intolerable. He is a boy of about twelve, with dreadful streaks of commonsense worthy of a man of fifty who has left all his illusions behind him. Yes, monster, that is you!"

Martin raised his eyebrows, his excellent temper slightly ruffled for the moment.

"Indeed, I didn't recognise it," he said.

"Dear Martin, don't be pompous. You didn't recognise it because it wasn't flattering. They say we women are vain, but compared to men— Some women are vain of their appearance, it is true, and usually without sufficient cause, but all men are vain of every attribute that God has or has not endowed them with. Remember that, Stella; and if you want to lead a quiet life, lay on flattery with a spade. They are insatiable. Personally I don't flatter Sunningdale, because I don't in the least want a quiet life. Tranquillity is so frightfully ageing, and makes one like an oyster."

Martin had recovered his serenity.

"When I am dead," he remarked, "you will be sorry for what you have said. But why this sudden attack on me?"

"When you are dead, you will see how right I was. But the attack—well, chiefly because you haven't provoked it. That is so tiresome of you. You could see I wanted to quarrel, and you wouldn't say anything I could lay hold of. If I want to sit down, politeness ordains that you should give me a chair. If you see I want to quarrel, politeness ordains that you should give me a pretext. It is the worst possible manners not to. My nerves are all on edge; when that is the case, the only thing to do is to quiet them by being rude to other people. Dearest Stella, you look too lovely this afternoon. Why you want to throw yourself away on Martin I can't think!"

"But you said just now it was your idea," said Stella.

"I know it was, and a very foolish one. I never imagined you would take it seriously. Besides, you know perfectly well that whenever a thing happens that pleases me, I always say it is my own idea. My darling, did I tread on you? How foolish of you to lie there! And when you are all happily settled for, what am I to do next?"

The clock struck and Martin looked up.

"Gracious, I am late," he said. "Karl was to give me a lesson at six. You must say good-bye to me next, Lady Sunningdale."

Stella got up too.

"I'll see you safely out of the house," she said, and left the room with him. Then, having closed the door, she paused, and taking hold of the lappel of his coat:

"Martin, you're not vexed with me?" she asked.

"No, why? I thought you were vexed with me."

"No, dear. I was vexed with myself, I think, and so I was horrid to you. But, my dearest, give me all you can of yourself. I want so much, just because it is you!"

Martin's eyes kindled and glowed.

"It is all yours," he said. "You know that. I wish there was more of it. And there is more since—since a week ago."

"Then I am content," she said, "and that means a great deal. I think I was rather jealous of pianos generally. And you forgive me? Yes?"

Lady Sunningdale, though often irrelevant from sheer irrelevancy, was also sometimes irrelevant on purpose, using preposterous conversation, as Bismarck used truth, as a valuable instrument to secure definite ends. Just now, for instance, her attack on Martin had purpose at its back, for she had seen quite distinctly that something had gone wrong between him and Stella, and had made the diversion in order to prevent the topic of friction, whatever it was, being subjected to further rubbing. Providence had lent aid to her benevolent scheme, sending Martin off to his music-lesson, and leaving Stella alone with her. In fact, her request to be told what she should do next needed no answer at all, for she knew quite well that what she would do next was to get Stella to confide in her, and tell her all that had happened. She was a great believer in talking things out; the important point, however, was not that the principals should talk things out, which was, indeed, worse than useless, but that they should severally talk it out with somebody else. She wondered, and, indeed, rather hoped, that Martin might simultaneously talk it out with Karl, for, as she had had occasion to observe before, Martin's music-lessons consisted chiefly of discussions on character.

Stella returned in a moment, and Lady Sunningdale was irrelevant no longer. She only took a preliminary circuit or two in the manner of a homing pigeon, before it takes the straight unswerving line.

"Martin is simply absorbed in the thought of his concert," she said. "And he is going to play just *all* the things that make me laugh and cry. Personally I shall go with five handkerchiefs and a copy of some English comic paper. The handkerchiefs are for the tears I shall shed, and the comic paper is to check my laughter when he plays the Paganini variations. Dear Stella, how very wise of you to marry a genius! You will never be dull. But it is rather bold too. Oh, please take Suez Canal out of the grate; he is trying to commit suicide, I think, because Sahara is not here. Yes. Geniuses are so unexpected and violent. It must be like marrying somebody who keeps several full-sized flashes of lightning about him, and also a large lump of damp clay. You never know which you will put your hand on, and they are both so dreadfully disconcerting."

Stella picked Suez Canal out of the grate. Apparently he was putting ashes on his head as a sign of mourning, and she dusted him carefully before replying.

"I am disconcerted," she said.

Lady Sunningdale never pressed for a confidence. "To show that you want a thing," she once said, "usually means that you are grudgingly given half of it. But if you firmly turn your baçk on it, it is hurled at you." She turned her back now, using irrelevance again.

"It is nearly three years since I was disconcerted," she said, "and the terrible thing is that I quite forget what disconcerted me. I think it must have been Sunningdale. Do you know he spoke in the House of Lords the other day on one side, and then voted on the other? His reason was that he felt his own remarks to be so feeble that he was sure there was more to be said on the other side. But I believe he merely forgot. Yes. That marble fireplace is so good. Surely it must be Adams'."

This was completely efficacious.

"Shall I bore you if I talk to you?" asked Stella.

"No, dearest Stella. I love being talked to. What is it?"

"It is Martin," said she.

The back view had done its part. Lady Sunningdale turned completely round again.

"Dearest Stella," she said, "pray put out the electric light. It is rather strong in my eyes. Yes, Martin, now!"

Stella felt as she turned out the light that this was exactly what she wished. In the dim flickering fire-light, her thoughts, drawn to the surface, became articulate more easily.

"He is just what you say," she said. "You touch him, and never know whether it is going to be lightning or clay. The lightning does not disconcert me. But, dear Lady Sunningdale, the clay does!"

Lady Sunningdale was really immensely interested. She had her own methods of getting the girl to rummage in the dark corners of her mind and bring out all that was there, and she pursued them now.

"Clay is not really disconcerting," she said. "It is only the possibility of clay when you expect lightning. My own darling Sunningdale is entirely clay. Of course there is clay in Martin: there is in everybody. How have you managed to come across it? Because he has singularly little."

"Music is his lightning," said Stella.

"Do you mean that the rest is clay?" asked Lady Sunningdale.

There was a pause, and Stella turned out an extremely dark corner in her mind: something really quite below the stairs.

"What if I am?" she asked.

"Then, dearest Stella, you have only yourself to thank. He did not think you clay anyhow a week ago. Else, why should he have asked you to marry him? Or do you mean that Martin has changed since then?"

Again Stella paused.

"I must say it more simply," she said. "Look at it in this way. What if Martin *is* music: if everything else to him is secondary to that?"

"Then he would have asked the complete works of Chopin to marry him," remarked Lady Sunningdale. "But as far as I know, he didn't. It occurs to me that he asked you. And I know, I can feel it, that he is devoted to you, really in love with you. Only don't for heaven's sake let your mind dwell for a moment on the relative positions that you and music hold to him."

"I have done worse than that," said Stella. "I have

asked him what relative positions we hold. I did so to-day."

"My dear, how insane! What did he say?"

"He told me not to talk nonsense. But is it nonsense?"

Lady Sunningdale drew a little nearer to the fire. All her kindliness, all her good nature, and, what was perhaps even more important, all her tact and finesse was enlisted on behalf of these two. She recognised to herself that there was here in all probability only one of those tiny misunderstandings which must occur between a man and a woman who are now for the first time really learning each other. At the same time it seemed to her quite important, if possible, to thoroughly dust, clean out and disinfect this dark little mental corner in Stella, for it might easily contain the germs of a misunderstanding that would be by no means trivial.

"Yes, it is nonsense," she said decidedly. "It is poisonous, suicidal nonsense. You are exactly like the Bear. You don't seem to grasp, any more than the Bear does, what music means to Martin. It means in one word 'God.' It is his religion—and good gracious, supposing he was a bishop and you were going to marry him, you would not, I hope, be jealous of his religion. And in music Martin is a very big bishop indeed! But in other respects—you forget this too—he is simply a child. I can't imagine what Martin will be like when he is middle-aged. It is impossible to think of him as middle-aged. Martin and middle-age are not compatible terms. True, Karl says he has been having a good many birthdays lately. I, too, think he has; but he has, so to speak, made saints' days of them all, and dedicated them to his religion. All but one, that is to say."

"And that one?" asked Stella.

"He had a birthday when he fell in love with you. That is yours: he has given you that. My dear, he adores you. When you come into the room his face is lit. Only for heaven's sake don't worry him and question him about his soul and his depth and the exact way in which he loves you. If you insist he will try to answer you, and his answers will be dreadfully disappointing to you, because he doesn't know anything about it. To question him is like—it is like looking at light through a prism or a spectroscope, splitting it up into rays, when instead you might be sitting in the sun. Dear me, how very precise and definite I am becoming! I mean exactly that—I hope I am not going to be ill."

Stella laughed.

"Dear Lady Sunningdale, I hope not," she said. "In any case tell me some more first."

"My dear, I can't talk sense to order. You must collect the extremely valuable grains of gold in my conversation for yourself, out of the extraordinary mass of quite valueless material."

"But he is disconcerting," began Stella again.

"Ah! yes, but so quite certainly are you to him. Heavens! how dull it would be if other people never disconcerted one! But I don't think Martin—though I am sure he must often find you disconcerting—would ever say so."

Stella flushed slightly.

"Is that a reproof?" she asked gently.

"It certainly is if it occurs to you that it may be, so pray, pray don't deserve it again. Where is Suez? Oh, there. And don't allow yourself, ever allow yourself, to think, 'What a pity there is an occasional lump of clay. For indeed there is so much lightning. If there wasn't a little clay, I really think Martin would explode, go off in spontaneous combustion. My dear, hours and hours of every day pass for Martin at a pressure of which stupid people like you and I have no conception. He recuperates by restful intervals, by being a mere boy with huge animal spirits. You may thank your stars he does not recuperate by being vicious or sulky. Most geniuses are morose, and very few are quite sane. Martin is quite sane, and even the Bear, who takes the gloomiest possible view of him, couldn't call him morose. Go down on your knees, my dear, and be thankful!"

Stella was silent for a moment. Then another corner was turned out.

"And there is no doubt about his genius?" she asked at length.

"But what is the matter with you?" asked Lady Sunningdale. "You will ask me next if I am quite sure he hasn't got false teeth. Dearest Stella, do drop this exacting, questioning attitude once and for all. I know almost everybody has an occasional attack of it, but, I am sure you will pardon me, it is just that which makes people odious. It turns them sour. For heaven's sake, don't turn sour. Suez Canal is in the grate again; oh, naughty! Thank you, dearest. Yes, sour. Take things on broad, indulgent lines. He loves you; that on the whole you believe to be a true statement of the case. Well then, surely that is good enough. Don't say, 'Does his love measure six feet in height, or is it only five feet eleven and three-quarters.' In fact, open the windows."

Stella took this very attentively and very gravely.

"Dear Lady Sunningdale," she said, "I am very grateful. I think you have done me good. I had a little attack of indigestion in my mind. Do you know, I never thought that you——"

"You never thought that I could think," said she, "and I'm sure I don't wonder. But I can think when I choose. Just now the object of my thought is to stop you thinking. Leave psychological questions alone when you are dealing with Martin. Just open your mouth, shut your eyes, and see what he will give you, as we used to say when children. You are a most fortunate girl. Heavens! fancy having Martin in love with one!"

There was the ring of absolute sincerity about this, so true and distinct that Stella wondered. She wondered still more, when, on looking at the other's face, she saw that Lady Sunningdale's eyes were full of tears, which she openly mopped up with a square two inches of lace.

"Yes, real tears," she said, "tears of extreme middle age, my dear. What are they made of? Water, I suppose, with just a little jealousy, and a little youth still left in them, and adoration for genius, and love of beauty. In fact, they are the most complicated tears I ever heard of, one or two like that from each eye and then it is over. Dearest Stella, you are such a fool. One is always a fool till one is middle-aged, and then one is young no longer. That is the tragedy of growing old. It is almost impossible to be mature and young simultaneously. You are a fool because you don't know what a priceless perfect gift has been given you, Martin's love. I envy you intolerably; I gnash my teeth with rage. Don't misunderstand me; I don't want him in the least to fall in love with me, and, to reassure you, I may say that even to my amorous eye there does not appear to be the very slightest chance of it. But I gnash my teeth, because I am not young like you, so that he might fall in love with me, and at the same time wise like myself, so that I should know what to do with him."

"Ah, tell me that, do tell me how to manage, how to behave," said Stella.

"I can't. That is just it. There is another tragedy in this mismanaged world, that nobody can teach anyone else anything that is worth knowing. You can't teach me how to look young: I can't teach you how to be wise, how to appreciate, how not to worry. But Martin's mind is like a cut diamond; it absorbs whatever light, blue, green, red, is thrown into it, and turns it by its own magic into inapproachable colour. That colour is seen in his music. Oh, I have watched you often this last week. You worry him and puzzle him, and I'm sure I don't wonder, if you ask him the relative places of music and you in his mind. Do you not see how stupid that is? Answer me."

Stella smiled.

"Oh, don't rub it in," she said. "Yes, it is idiotic."

"My dear, you are so gentle that I feel a brute!"

"Please be a brute then, just five minutes more," said Stella.

"Very good. Do not take up this absurd position and say, 'I am your goddess; what incense have you got to burn before me this morning? Ah, that is the second quality incense! I thought so. How *could* you?' Be much bigger than that. Suez! Recollect who it is who has paid you this incomparable compliment of saying he wishes to see your face opposite him at breakfast for the rest of his life—every day, every day. Go to Karl Rusoff and ask him where he places Martin, if you do not believe me about his genius. And when he has told you, hire the Albert Hall, fill it with people, and tell them what Karl says. Then wait a couple of years, hire the Albert Hall again, and repeat again what Karl told you. And every single person in the hall will say, 'Why, of course. We knew that.'"

Stella was silent a moment.

"Then must I burn incense before him?" she asked. "The very best incense? I should love to do that!"

Lady Sunningdale restrained a movement of impatience.

"My dear, you are the one person in the world who must not burn incense," she said. "An incense-burning wife is like dram-drinking to a man. You are to be his wife. That means a good deal. But you are to be his comrade. That means much, much more. He and Helen! Why he did not get Helen to come and live with him, and—well, not marry at all, I don't know. Perhaps Frank would object. Men are so selfish."

"Do you mean he has chosen badly?" asked Stella.

"No, dear, and it is silly to say that. What I meant was that I wonder why he wanted to marry at all, why a nature like that has need of anybody else. If I was like Martin I should never see a soul, but contemplate my own wonderfulness. However, he did want somebody else. And he chose you, you fortunate girl."

"I ought to be very happy, then?" she asked.

"Ah, I don't say that. Perhaps you will be divinely, ecstatically discontented. Happiness is rather a bovine quality, I always think. It implies not wanting. Anyone with imagination must always want. Yes. Dear me, I came here to say something, and I forget what—I have said a good deal, but not it. Dearest Stella, do you forgive me? At least for my own creature comfort I want you to forgive me: but essentially I don't care, as I know I am right."

"No, I don't forgive you," said the girl, "but I thank you."

Lady Sunningdale struggled to her feet out of her very low chair.

"That is sweet of you. Yes, Suez, my darling, we are going home to din-din and Sahara. Ah, I remember. I want you and your mother to join us at Cannes for a fortnight at Easter. Sunningdale's villa is really quite comfortable, and you can look at the Mediterranean and meditate. Ask her to send me a line about it, but come yourself in any case. The southern sun always melts my brains, and liquid wisdom flows from my lips in practically unlimited quantities. Why don't we all live at Cannes, among the palms and that sort of thing? If you can't come I shall ask Martin, but I don't mean to have you together. You will be quite enough together afterwards. Dear me, how screaming Martin will be as the master of a house! Good-bye, darling Stella. Yes, pray turn up the lights, otherwise I shall crash my way through priceless furniture, and tread on Suez Canal."

CHAPTER XIV.

KARL RUSOFF had experienced a good deal of inward anxiety, which he was very careful to keep entirely to himself, for several days before Martin's concert, for the thought of it, as the day got near, had agitated and excited the latter to the point of making him lose his sleep and his appetite. Though Karl knew quite well that an artist does his best as a rule under the spell of excitement, more, that any notable achievement can hardly be compassed without it, yet in the present case, Martin himself was naturally so highly strung, and his excitement had become acute so many hours before he was to make his appearance, that his master could not help silently wondering whether he could stand the strain of it till the day came. At other times again, Karl, knowing Martin's serene splendid health, found consolation in telling himself that the tighter and more tense his nerves got, the more wonderful would his playing be. Even during the last week or two he had made such an enormous advance in his general grasp, that Karl knew that he himself would be bitterly disappointed if this extraordinary youth did not on his very first appearance legitimately and justifiably take musical London by storm. At the same time he knew that he himself would give a very deep sigh of relief when Martin had got through, say, the first three minutes of his recital. That safely past, he was sure that the mere feel of the familiar notes would occupy him to the exclusion of all agitation.

Only a quarter of an hour before he was to come onto the platform, Karl was with him in the artists' room, trying to occupy his mind in talk, but watching him with ever increasing nervousness, as he walked up and down like a caged animal between door and window. Once Martin took out a cigarette, bit the end off as if it were a cigar, and threw it away. Then he asked a question, paid not the slightest attention to the answer, and finally sat down on the edge of the table. His face was flushed, his eyes very bright; had not it been that Karl knew how excited he was, he would have thought he was ill.

"I shall break down," he said; "look at my hands, look how they tremble. I can't keep them still. I could no more play a series of octaves than I could fly. It would be like the 'Tremolo' stop on Chartries organ."

"My dear boy, I have told you that does not matter in the slightest degree," said Karl. "The moment you touch the notes that will cease absolutely. Why, even now my hands tremble before I begin!"

Martin apparently was not listening.

"And I have not the remotest notion how the 'Études Symphoniques' begin," he said.

Karl tried to laugh, not very successfully. As a matter of fact he was quite as nervous as Martin.

"That's a great pity," he said, "as you open with it. I don't know either."

But Martin did not smile.

"What will you do if I break down?" he said; "if I can't begin? It is more than possible."

"I shall hiss; I shall boo; I shall demand the return of my money," said he.

But Martin still remained perfectly grave.

"Ah, don't," he said; "the others may boo if they

like, and I shan't mind—much. But I couldn't stand it if you did."

"Did you drink a good stiff glass of whisky and soda for lunch, as I told you to?" demanded Karl.

"I tried to, but I should have been dead drunk if I had gone on. So what will you do if I break down?" he asked again. "You told me, but I have forgotten."

Karl rose from his chair.

"I shall break my heart, Martin," he said.

Then he spoke to him quickly, peremptorily, seeing he was really on the verge of hysterics.

"We've had quite enough of this nonsense, my dear boy," he said. "If you give me any more of it, I shall lose my temper with you. You are not going to break down; I forbid you, and you are to do as I tell you. You are going to play your very best, better than you have ever played before. Now I must get to my place. Give them five minutes' law before you appear, and as soon as I see the top of your black head coming up the stairs, I shall have all the doors closed till the end of the 'Études.' We'll have no interruptions; they are frightfully distracting. You know where I shall be sitting, don't you? Bow twice, right and left, walk straight to the piano, and begin instantly without playing any fluffy arpeggios. It is going to be a great day for you—and for me."

Martin looked despairingly round.

"Don't leave me, don't leave me," he said. "Can't you sit by me?"

"And hold your hand? Ah, this is altogether childish!"

For the first time the shadow of a smile crossed Martin's face.

"I know it is," he said. "I can just, just see that. I think I had better try to be a little man for a change."

The hall was crammed to overflowing, as if some pianist of world-wide fame was to make his appearance, and not a young man who had never appeared in public before. Several causes had contributed to this, the first and most important being that Mr. Martin Challoner was actually a pupil of Karl Rusoff's, who for years had never consented to teach. Furthermore, Karl Rusoff had the very highest opinion of him—exaggeratedly high perhaps, since he was his pupil—and had not only allowed, but wished him to give a concert. Surely then he would run no risks: Martin Challoner must have some merit. In addition, no English pianist of more than mediocre powers had appeared for years, and patriotism called. Finally for the last fortnight Lady Sunningdale had worn her coachman to a shadow and her horses to skin and bones, so incessantly and unintermittently had she driven about, first of all to the houses of her intimates, then of her ordinary friends, and finally of the merest acquaintances, practically insisting that they should all appear. Karl Rusoff had done what he could to discourage this, but his efforts were totally void of effect, for Lady Sunningdale had told him that it was her "duty" to do her best for Martin. She seldom used the word "duty," but when she did it might be defined as anything she was irrevocably determined to do, from which no argument could move her.

So, for the first time, Martin found himself in that unspeakable position of being alone on the shore of a sea of faces, the owners of which had paid money in anticipation of the pleasure he had undertaken to provide for them. Opposite him, a few yards off only, but looking misty and unreal, was the Steinway grand, and he found himself wondering what on earth it was for. When he remembered, he felt towards it as a condemned man may feel when he sees the execution shed at a few minutes before eight. Then he bowed in answer to a very fair reception, and walked straight to the piano. He glanced at his programme, and saw he had to begin with Schumann's "Études Symphoniques." He sat down, waited a moment for silence, and began.

He played one bar only and then stopped. He had not the very faintest idea of how it went on, and in a sort of mild despair—he felt as if his powers of feeling were packed in cotton wool—looked down to where Karl was sitting in the third row. Those great grey eyes were fixed on him with an expression of supreme appeal; he could see the master's hands clutching convulsively at the back of the seat in front of him. And at that sight, at the sight of the agony Karl was in, Martin was able for one moment to forget himself and all the bewildering crowd of faces. So, fighting against the paralysis that was on him, no longer for his own sake, but for Karl's, he again turned to the piano.

But still he could think of nothing, nothing; he could not even remember the first bar that he had played just now, and he bit his lip with his teeth till the blood came, saying to himself, "It will break his heart; it will break his heart." The numb dulled sense was gone; in that half minute he endured an agony of years.

Then quite suddenly, like the passage of the sun from behind some black cloud, all came back to him, and he sat still a moment longer in sheer happiness. At the concentrated thought of what Karl was suffering, his nervousness, his paralysis of mind went entirely from him; and with complete certainty, with the assured knowledge too that he was going to play his very best, he began again.

At the end of the slow thema he paused, looked up at Karl, and smiled nearly to laughing-point at him, pushed back the plume of hair that dropped over his forehead and—played. And at that smile and at the gesture that was frequent with him, Karl gave one immense sigh of relief that Martin could hear. But now it meant nothing to him; he was busy.

Martin's face during those few horrible moments had grown absolutely colourless, so that Karl had thought, and almost wished—for so the public shame would be lessened, and people would be compassionate—that he was going to faint. For when for the second time Martin had turned to the piano, and still could not begin, he believed for that moment that the boy could not pull himself together, that unless he fainted, he would simply have to walk off the platform again. But now the colour came back, slowly at first, then with sudden flushes: the dead apathy of his face changed, and began to live again. Soon his mouth parted slightly, as if wondering at the magic of the music which blossomed like roses underneath his flying fingers. Once or twice between the variations, he brushed back his hair again, once he looked up again at Karl with the brilliant glance his master knew and loved, asking with his eyes, "Will that do? Will that do for *you?*" before he went on interpreting to the breathless crowd the noble joy which must have filled the composer as he wrote. Full of artistic triumph as Karl's life had been, never before had it mounted and soared so high as now, when not he, but his pupil held the hall enchained.

And in that moment his own ambitions which he had so splendidly realised for so long, dropped dead. He and Martin, he knew now, were master and pupil no longer; it was the master's turn—and with what solemn joy he did it—to sit and learn, to hear—and he longed for a myriad ears—what was possible, for even Martin had never played like that before. Even admiration was dead: there was no room for anything except listening: admiration, wonder, delight, laughter of joy might come when the last note had sounded, but at present to listen was enough.

Martin held the last chord long. Then he took both hands off, as if the keys were hot, and rose, facing the hall. For him too, just then, personal ambition was dead; he had played, as David played before Saul, in order to drive from his master's face the demon of agony that he had seen there. And he looked not at Stella, not at Lady Sunningdale, not at Frank and Helen, nor did his eyes wander over the crowded rows, but straight at Karl, while the hall grew louder and louder, till the air was thick with sound, still asking him, "Did I play it well?" And when Karl nodded to him, he was content, and bowed in front of him and to right and left, thinking "How kind they all are!" He caught Stella's eye and smiled, Frank's, Helen's, Lady Sunningdale's. Then he sat down at the piano again.

But it was quite impossible to begin, and for his own amusement (for now, it must be confessed, he was enjoying himself quite enormously) he struck an octave rather sharply, and heard not the faintest vibration from the strings above the uproar. So he rose again, bowed again, and still bowed, and bowed still, till he felt like a Chinese mandarin, and knew everybody must think so too. Then he sat down and waited till the phlegmatic English public had said "thank you" enough.

A ten minutes' interval had been put down on the programme, and tea was waiting for him in the room below. But he forgot all about it, and went straight through. The recital was carefully chosen not to be too long, and in the ordinary course of events the audience would have been streaming out into the street again after an hour and a half. But they refused to stream: Martin gave one *encore,* and after a pause a second, but he was still wildly recalled. Once before in the summer he and Helen had sent London mad about them; this afternoon he did it alone. And at last in a despair that was wholly delightful, as the hush fell on the house again when he sat down for the fourth time, he played "God save the King" solemnly through, and his audience laughed and departed.

Lady Sunningdale found that she had burst her left-hand glove, and lost her right-foot shoe, when she came to take stock of what had happened, as Martin finally retired after "God save the King." Karl was sitting next her.

"Don't speak to me anybody," she said, "because there is nothing whatever to say. That is Martin. I knew it all along. Yes, a shoe, so tiresome, I don't know how it happens. Thank you, Monsieur Rusoff. Stella, dear, we start from Victoria to-morrow morning, not Charing Cross. What did I tell you when we talked last? Do you not see? That is Martin. If anyone speaks to me I shall slap him in the face and burst into floods of tears. I should like to see that darling for one moment, just to tell him that he has not been altogether a failure. Which is the way? I suppose he is drinking porter now, is he not, or is it only singers who do that? Eight o'clock, Stella. Quarter to eight, Frank, because you are always late. Dearest Helen, how is the Bear? Yet Martin has only got eight fingers and two thumbs like the rest of us. And was it not too thrilling at the beginning? I knew exactly how he felt. It was pure toss-up for just one moment whether he would be able to play at all, or send us empty away, like the 'Magnificat.' Through this door, isn't it?"

Karl Rusoff showed her the way through the short passage into the room where two hours ago he had sat with Martin on the verge of hysterics. But now a great shout of boyish laughter hailed them, and Martin went up to Karl, both hands outstretched.

"Ah, it was you who pulled me through," he said. "I couldn't have begun otherwise. But it hurt you so dreadfully. I—I felt it hurt you. And shall I ever play like that again? I never played like it before!"

Karl looked at him a moment without speaking. Then he raised the boy's hands to his lips and kissed them.

"I mean that," he said. "Ah, Martin, how I mean that!"

Martin stood quite still. Had such a thing ever suggested itself as possible to him, he would have felt ready to sink into the earth with sheer embarrassment. But now when the unimagined, the impossible had happened, he felt no embarrassment at all.

"You did it all," he said simply. "Thank you a hundred thousand times."

Then the pendulum swung back again, and he was a boy, and boyishly delighted with success.

"Oh, I enjoyed it all so," he said. "After that first terrible minute I just revelled in it. Can't I give another concert this evening?"

Here Lady Sunningdale broke in—

"You not only can, but you must after dinner," she said. "Martin, you played really nicely to-day. I am going to begin to practise to-morrow morning. Scales. No, not to-morrow morning, because I shall be otherwise engaged on the English Channel. Why can't they run a large steam-roller over the sea between Dover and Calais? Nobody can tell me. However, I'm told it is rather healthy than otherwise. My dear, red velvet sofas, tin basins, Stella, and I. Also Suez Canal. Sahara is not yet in a fit state. It is too terrible. Eight o'clock to-night, Martin. And I shall never forgive you for this afternoon. You gave me the worst five minutes I ever had."

"I tried to make up for it," said he.

Lady Sunningdale turned back in the doorway.

"I adored you," she said. "And next time I shall wear large eights. Perhaps they will not burst quite so soon."

Martin turned a thirsty eye on Karl when she had gone.

"And can I have my whisky and soda now?" he asked. "I want it frightfully."

Then quite suddenly his face changed, as if a lamp had been put out. He looked tired, worn out.

"And I have such a headache," he said. "I think I have had it two days, but I was too excited to think about it. It went away altogether when I was playing. But it has come back in force."

Karl rang the bell.

"Yes, you want a good rest," he said, "you are tired without knowing it: you have been living on your nerves the last day or two. But anything worth doing is worth being tired over. Dear boy, I hope your headache is not really bad. Anyhow you have done the thing worth doing. Don't go out to-night. Go back home, and go to bed early."

Martin shook his head smiling.

"Ah, I won't give up an hour of to-day for fifty headaches," he said. "Besides, Stella and Lady Sunningdale leave to-morrow. My father was not at the concert, I suppose?"

"No, not that I know of."

"I sent him a ticket, though I thought he would not come. He does not even now approve of my wasting my time at the piano," he added with an irritability to which this horrible stabbing pain in his head contributed.

He drank his whisky and soda with feverish thirst.

"And I had better have left that unsaid," he remarked. "Now I shall go home, I think, and sleep off my headache before dinner. But I must just look at the stage once more."

He ran up the steps, and looked round the empty hall. The lights were being extinguished, and gangway carpets being rolled up. The Steinway still stood there, and he felt somehow as if he were saying good-bye to it.

"Well, that is done," he said to himself.

Lady Sunningdale and Stella left London for the Riviera next morning, and later in the day Martin went down to his uncle's at Chartries, and Helen back home to the Vicarage. The reaction from the excitement of the last few days had left him, naturally enough, rather indolent and tired, and also, naturally enough, rather irritable and disposed—not to put too fine a point on it—to be cross. He found the railway carriage insufferably hot, and pulled down a window: that, however, made it draughty, and he changed his seat, and sat with his back to the engine. This, however, was no good, because for some unexplained reason it made him feel ill, and, changing back once more, he fell into a heavy sleep that lasted till they got to their station. Even then the stopping of the train did not arouse him, and Helen had to shake and poke him into consciousness, for which kind office she got growled at.

But he had come to Chartries with the definite object of seeing his father, and while Helen's luggage was being put into the pony-cart from the Vicarage the two talked this over.

"It's no use putting it off," he said, "so will you tell father that unless I hear from him to stop me, I will come over to-morrow afternoon to see him. And I hope," he added with his usual candour, "that my temper will be a little improved by then. Lord, how cross I feel! And this time yesterday I was in the middle of it all."

Helen looked at him a moment rather anxiously.

"You're all right, aren't you, Martin?" she said, "not ill?"

"Ill? No. But I'm all on edge, and I've got two headaches. It's rather cold waiting here. I think I'll walk on and let the carriage catch me up. Good-bye, Helen: see you to-morrow."

Martin woke next morning after long heavy sleep with the same sense of lassitude and tiredness which had oppressed him all the day before, and the same headache lying like a hot metallic lump inside his head, pressing the back of his eyes. The man who called him had brought him a couple of letters and a note from his father which had been sent over from the Vicarage. He opened this first.

"My Dear Martin,

"Helen has given me your message, that you wish to see me. I have thought about it very carefully, and I wish to tell you quite candidly the conclusion I have come to.

"You know what I felt about your going over to the Roman Church: I feel all that still, and as strongly as ever. You have deliberately left your own church, and for reasons, as far as I can understand, which are frivolous and unessential. And I am afraid—I know, in fact—that if I saw you, I should, without being able to help myself, express to you what I feel. Now I do not think this would do any good, it would only widen the gulf between us, and one of the great aims of my life now is to do the opposite. I do not suppose my opinion will ever change, it cannot in fact, but in time I shall, I suppose, get more used to what has happened, and shall be able to see you without bitterness. At present I am unwilling to tear open a wound which may be beginning to heal. But all this is to me still so keen a daily and hourly pain that I feel sure we should be wiser not to meet yet. But Helen of course is quite free to come and see you, and you to come and see her.

"It gives me great pain to write this. But I cannot separate you from what you have done.

"I am rejoiced to hear from her of the great success of your concert. Personally, as you know, I have no educated taste in music, but I gather that your master is satisfied both with your progress and your industry, which is more important than success.

"My dear boy, I wish I could see you: I wish I could trust myself!

"Your affectionate father,

"SIDNEY CHALLONER.

"P.S.—Your Aunt Clara, I am sorry to say, is in bed with a sharp attack of influenza."

Martin read this through twice before he got up: then he dressed, his cold bath making him shiver, and went downstairs. The sight of his own face in the looking-glass as he brushed his hair, was somehow rather a shock to him: it did not look exactly ill, but it was unfamiliar, it looked like the face of somebody else. His uncle was not yet down, and he strolled out onto the terrace, waiting for him, into the warm windy sunshine of the April morning. But here again he had the same impression of unfamiliarity: the sun did not feel to him the same, nor did the sunshine look the same, both light and colour had an odd dreamlike unreality about them. It was as if some curious hard barrier had been put up between his sense of perception and that which he perceived. Then with a feeling of relief he remembered his father's postscript. Probably he had influenza too.

That explanation, or the divine freshness of the morning, made him feel rather better, and half laughing at himself for his vague fear that there was something really wrong with him, he went indoors again. People were coming to stay at Chartries that afternoon, but this morning he and his uncle were alone. Lord Flintshire was already seated at breakfast when he came in.

He gave him his father's letter to read, unconscious that his uncle looked rather closely at him as he entered, being also struck by a curious drawn look in his face, but he said nothing on the subject, and read the letter through.

"I think your father is wrong about it," he said, "and if you approve, I will tell him so. There is surely no need to enter into theological discussion; you want just to see him, and shake hands with him."

Martin had taken some fish, but gave it up as a bad job, and drank tea instead.

"Yes, just that," he said. "I hate being on bad terms with anybody, specially him."

Lord Flintshire looked at him again.

"The boy's ill," he said to himself. Then aloud:

"Well, let us walk over after breakfast, if you feel inclined. You can see Helen, while I go in and talk to your father. You don't look particularly fit this morning, Martin. Anything wrong?"

"I feel beastly," said Martin with directness. "I shouldn't wonder if I had got influenza too."

"Are you sure you feel up to walking? Yes, your father mentions that Clara has got it. If the doctor is there, he might just have a look at you. Or if you don't feel up to coming, I would send him back here."

Martin pulled himself together: the tea had made him feel quite distinctly better.

"Oh no, I'm quite up to it," he said. "Probably the doctor will tell me to go for a long walk, and eat a big dinner. And I should like to see my father as soon as possible, and get it over. It will all be easier after that."

His uncle got up.

"Shall we start in half an hour, then? We shall be sure to catch him before he goes out. Cigarette?"

"No, I think not, thanks," said Martin.

Their way lay down through the woods where Helen and Frank had met a month ago, and the gracious influence of springtime had gone steadily forward with the great yearly miracle of the renewal of life. The green that had then hung mistlike round the trees was now formed and definite leaf, exquisitely tender and clear, and in this early morning hour shining with the moisture and dews of night. Daffodils still lingered in sheltered places and the delicate wood-anemone flushed faintly in the thickets. Below, the chalk-stream, where Martin last summer had spent that hour of self-revelation, was brimful from bank to bank of hurrying translucent water, which combed the sub-aqueous weeds, and turned to topazes the yellow pebbles and into heaps of pearl the beds of chalk that flashed beneath the water. But this morning he was heavy-eyed and clogged of brain; he felt that somebody else was seeing these things, that somebody else was putting foot in front of foot, while he himself had dwindled to a mere pin-point set in the centre of a great lump of hot metal which filled his head. Sometimes this body that was once his felt sudden flushes of heat, sometimes it shivered for no reason. Then after an interminable walk, so it seemed to him, they turned through the churchyard and went up the gravel path that ran to join the carriage sweep in front of the vicarage door. And in spite of all, it was with a wonderful sense of coming home that Martin saw the grey creeper-covered walls again, the long box-hedge, and the croquet lawn wet and shining with dew in the sun.

"I'll wait out here while you see my father," said he. "Perhaps you would tell Helen I am here." And he sat down all of a heap on a garden seat.

This tired, spiritless boy was so utterly unlike Martin, that his uncle felt suddenly anxious.

"Are you feeling bad, Martin?" he asked. "Do you feel faint? Hadn't you better come indoors?"

"Oh, no. I shall be better when I've rested a minute. But my head aches so. Lord, it gets worse every minute."

Lord Flintshire left him and went straight to Mr. Challoner's study, where he was at work.

"Good morning, Sidney," he said, "I have come over with Martin, who wants to see you. I also want you to see him, but we can talk of that afterwards. Now, is the doctor in the house? Martin is not at all well. He looks to me very ill. He——"

But at that word there was no longer any thought of "talking of that afterwards." All that was human and tender, all that was loving, all that there was of "father" in Mr. Challoner sprang to that call.

"Dear lad, where is he?" he said. "Yes, the doctor is with Clara now. He will be out in a minute. But where is Martin? I must go to him."

Lord Flintshire just laid his hand on his brother's shoulder.

"I knew you would, Sidney," he said. "He is outside by the front door."

Martin had dropped heavily onto the garden seat, and sat there with his eyes closed. That lump of hot metal in his head had grown larger and hotter; he felt as if something must burst. And he was so terribly tired; his walk had not done him the least good. Then he heard quick steps behind him on the gravel, but simply could not be troubled to look round. And then came his father's voice.

"My dearest lad," he said, "come indoors at once."

Martin sat up with a jerk, and some chord of old memory twanged on the surface of his brain.

"You're not angry with me, father?" he said nervously.

Mr. Challoner bit his lip to stifle the exclamation of pain that rose bitterly within him.

"Angry?" he said. "What put that into your dear old head? There, Martin, take my arm, and lean on me. Come inside out of the wind. There, old boy, steadily; there's plenty of time. I hope we shan't have you down with influenza too. But it's the luckiest thing in the world. The doctor is here now with your aunt, and he shall have a look at you."

But it needed all Mr. Challoner's courage to get through with this cheerful chattering; Martin looked terribly ill to him. But he got him into his study, arranged the cushions on the sofa he so seldom used himself, and made him lie down.

"Ah, that's better," said Martin. "Thanks, thanks ever so much, father."

He held out his hand to his father, who pressed it, and his voice trembled a little as he answered.

"God bless you, my dear lad, for wanting to come and see me," he said. "Now is there anything you want? I shall send Dr. Thaxter to you as soon as he leaves your aunt."

Dr. Thaxter was a merry, rosy-faced little man with a manner so reassuring that one felt quite well directly, and in a few minutes he came bustling into the room.

"Ah, Mr. Challoner," he said, "your father tells me you are a bit knocked up. Not uncommon in this spring weather. Quite right to lie down. There, put that under your tongue, and don't bite it."

He adjusted the thermometer and went chattering on.

"And you've walked over from Chartries with your uncle, have you? Fine place that, and a fine healthy situation. Of course you only came down yesterday. I saw the account of your concert in the paper. Ah, I wish I had been there. Now I think we've given the thermometer long enough. Thank you. And you feel rather——"

The little doctor stopped suddenly in the middle of his sentence, when he saw what was recorded on it.

"You have a headache, I think your father said."

"I have nothing else, I think," said Martin.

Dr. Thaxter drew a chair close to the sofa, and sat down, looking at him very closely.

"Ah, yes, that is to be expected with a little fever. You are rather feverish. Now when did you begin to feel ill? When did you first feel a headache? Try to tell me all about it."

"Oh, five days ago now. No, six, I think. I don't think I felt anything else, except that everything seemed rather queer all the time."

He made a movement to sit up, but the doctor gently pressed him back again.

"Better not sit up," he said. "You'll be far more comfortable lying down. And you can tell me nothing else? Just a bad headache."

"Am I ill?" asked Martin suddenly. "Really ill, I mean? What's the matter with me?"

"My dear Mr. Challoner, I can't possibly tell you, because I don't know. And when one doesn't know, one takes precautions against anything that it may conceivably be. Perhaps it is influenza; if it is, it's a pretty sharp attack. I wonder at your being able to walk over this morning. Now will you promise me to lie quite still, while I just go and talk to your father, and settle with him what we shall do with you."

The little doctor went quietly out of the room, and across the hall to the drawing-room. Helen, her father, and Lord Flintshire were all there. Dr. Thaxter did not look quite so brisk and cheerful as he had done before he saw Martin.

"He has a very high temperature," he said, "much higher than I like. It may, of course, be an attack of influenza; I have seen cases of it with temperatures higher than that. But he must be nursed as if something more serious was the matter. He has probably had a temperature for nearly a week."

Mr. Challoner turned to him almost fiercely.

"What is it?" he said.

"It may be several things. Perhaps I can tell you when I have seen him again, when we have got him to bed. Now is there a good spare room in this house?"

"Yes, his own," said Helen.

"Very well; he must be moved there, just as he is

without getting up. If you and Lord Flintshire will help me, we will do it at once. And is there a room where a nurse can sleep?"

Helen took a step nearer him.

"Is it typhoid?" she asked.

"I am afraid it may be. It looks very like it."

CHAPTER XV.

It was very early, only a little after six, and the sun had risen on a day exquisite, warm and windless. In Martin's room the big window had been open all night, and all night the blind had not once rattled or stirred, while the lamp on the table near it burned steady without a flicker. But though it had been light for nearly an hour, the nurse had only this moment put out the lamp, for she had been alert, quick, and watchful, unable to leave his bedside for a moment for the last four hours.

He had been very restless, attempting again and again to sit up in bed, and it had needed not only all her care, but all her strength to keep him lying down. All night long, too, that terrible uncontrollable twitching of the muscles of leg and arm had gone on incessantly, and again and again for ten minutes or more at a stretch she had kept one arm with steady pressure over those poor jumping knees, while she held the other ready to prevent his getting up. It had been all she could do in fact to manage him alone, but she had been unwilling, except at the last extremity, to rouse Nurse James from the next room, for she had had a terribly tiring day yesterday with him. Yesterday, too, a second doctor had come down from London; the case was extremely grave, but all that could be done was being done.

Martin was lying rather more quiet just now, and Nurse Baker had moved from the bed to put out the lamp, and draw the blind up a little. His eyes were wide open, staring at the ceiling, and he was talking in a high meaningless drone.

"No, Karl, I can't do it," he was saying. "I don't see it like that. I know I shall break down, because I haven't the slightest idea of how it begins, and I can't leave out the beginning. And father is angry with me, and when he is angry he frightens me. Hasn't Stella come to see me? I had such a headache, you know, like a great piece of hot iron right inside my head. They took off the top of my head to put it there. I'm frightened of him when he's like that. Where's Stella? No, Lady Sunningdale was in the bird of para—para—parachute. I don't know; in that hat, anyhow, you fool, with Sahara. That's what made it so hot, and I can't endure English chants. Oh, father, don't, don't. It isn't my fault."

His voice rose to a scream, and the nurse came quickly back to the bedside, just in time to prevent him rising.

The door opened gently, and Helen came in, in her dressing-gown. And the terrible drone began again.

"And when we're married, Helen and Frank shall come and stay with us, and I'll play to them, if it gets cooler. But father mustn't know; he mustn't come. Karl is the loud pedal, you see, and the music stool, and I'm only the black notes. I hope they won't play me much, as I'm all out of tune with the iron. And all those faces are there, a sea of them, and I'm all alone. If I break down, father will be angry!"

He turned his head sideways onto the pillow, closed his eyes, and was silent for a little. Helen, with quivering lip, was looking at that dear face, so thin and hollow, so untidy and unshaven, with unspeakable love and longing. Then the nurse left the bed, and came to her. Helen did not ask if he was better.

"Can I help you in anything?" she said.

"No, dear Miss Helen, thank you. I think he will be quieter for a little now. But I should like Dr. Thaxter to be sent for at once, please. Yes, he is very ill. He is as ill as he can be. There, there, my dear."

Helen clasped her hands together a moment, holding them out towards Martin with a dumb, beseeching gesture, as if imploring him.

"And I am so strong," she said. "Why can't I give him some of my strength? It is cruel."

"Ah! if one only could do that," said Nurse Baker. "But he is not suffering; he is quite unconscious."

"May my father come in to see him a moment?" asked the girl.

"No, much better not. He does not know what he is saying, but he keeps on saying what you have heard. Now will you send somebody for the doctor? There are certain things I don't like about his looks. And then come back, dear, if you like. He never says a word his sister should not hear."

Helen advanced to the side of the bed for a moment,

and just touched Martin's hand, which lay outside the bedclothes. She could not speak, but just nodded to the nurse and went away.

She sent word to the stables that the cart was to go at once to fetch Doctor Thaxter, and then went to her father's study, where he was waiting for her.

He was kneeling by his table, as he had knelt for the last half-hour, but rose when she entered, and they stood together, hands clasped a moment.

"No, dear father, he is no better," she said. "He—he is very ill indeed. And Nurse Baker thinks you had better not go in."

Mr. Challoner looked at her with that dreadful dry-eyed despair that she had seen on his face so often during this last week.

"Does he still talk about me," he asked.

Helen laid her hands on his shoulders.

"Yes, father," she said, "but he does not know what he is saying. Indeed he does not. He talks all sorts of nonsense. He has no idea what he says."

"Ah! Helen, that is just it," he moaned. "The poor lad speaks instinctively; he says what has become a habit of thought. Oh, my God, my God!"

Helen knew her impotence to help him.

"I have sent for Doctor Thaxter," she said. "Nurse Baker wanted him to come at once. And, father, there is another thing, which I have only just thought of. If Dr. Thaxter thinks—if he thinks *that,* we ought to send for a Roman priest."

Mr. Challoner's face changed suddenly.

"No," he said in a harsh whisper, "no Roman priest shall enter the house."

"Ah! but he must, he must," said Helen. "Think a moment. If Martin was conscious, you know he would wish it, and you would send for one."

Mr. Challoner did not reply for a moment; then he lifted his hands with a helpless gesture.

"And it is Easter morning," he said.

Somehow that cut at the girl's heart more than anything.

"Yes, dear father," she said at length, "and is not that—whatever happens—enough for us all? Whoever we are—Frank, Martin, you, I—that is where we meet."

Then for the first time since that day now nearly a fortnight ago, when Martin had sat down dead-tired on the seat by the front door, the blessed relief of tears came to his father, and he wept long, silently, a man's hard, painful tears. And with those tears the upright hardness of him, the God-fearing, God-loving narrowness went from him. The bitter frosts of his nature melted, they were dissolved.

"Oh, Helen, if he lives," he said at length.

"Ah! yes, dear father, or if he dies. Even if he dies, dear."

She took his hands, holding them tightly.

"Oh, help me to remember that," she whispered, "I shall need all the help you can give me. We shall want—we shall want all the help we can get—both of us. We will give it each other. And Stella——"

"You telegraphed to her?"

"Yes: she cannot get here till to-morrow!"

Then the girl gave way.

"To-morrow," she said, "and it is only just to-day. Father, father. I can't bear it. I can't."

But the strength she had given him so often during this last week was ready again to help her.

"Yes, dear Helen," he said, speaking quite calmly again. "We can both bear whatever is to be. God does not send us anything that we are not capable of bearing, and of bearing without bitterness and without complaint. And whether it is life or death with our dear Martin, it is all life. We believe that, do we not? Let us hold on to that, for it sustains the sorrows of all the world. There is nothing so sure as that. It is the Rock of Ages, Helen."

There was the sound of wheels on the gravel outside.

"That will be the doctor, dear," said he, "will you go and meet him, and—and the cart must wait if he thinks a priest should be sent for."

She got up at once.

"Yes, father," she said.

Helen went out into the hall. Dr. Thaxter had just come in, and at the same moment Nurse Baker hurried downstairs.

"Come up at once, please, doctor," she said. "He—he came to himself a few minutes ago, after being delirious all night. I took his temperature. It is normal, just about normal."

Helen's face suddenly brightened.

"He is better, then?" she said.

Nurse Baker turned to her as the doctor took off his coat, with infinite compassion in her kind brown eyes.

"No, dear Miss Helen," she said. "He is—ah, I need not explain to you. But it is very bad. It is

—you must be very brave, my dear. Go to your father."

She gave her a quick little kiss, and followed the doctor upstairs. Helen went back into the study.

"Something has happened," she said. "I had no time to speak to Dr. Thaxter. They will send for us, dear. I think—I think that is what nurse meant."

It was now about seven of the morning, and the sun, an hour above the horizon, streamed gloriously into the room. It shone on the table, the sofa, on the big chair where Helen and Martin as little children used to sit together, looking at Bible pictures. And she sat down in that chair now. The big things had been said between her father and her, and, as they waited now, both turned to little memories of the past.

"Martin used to sit by me," she said.

"Yes, and then you grew too big. After that you used each to have a chair, one on each side of me."

"And we did our lessons there," said Helen. Then she stopped suddenly, for there was a foot on the stairs.

Nurse Baker came in.

"You must both come," she said.

The blind was drawn up in Martin's room, the same wonderful sun flooded the floor, and outside many thrushes were singing. There was but little apparatus of medicine there, it was just a boy's clean room: cricket-bats and racquets stood in one corner, on the table there was a heap of music, school-books were in the bookcase by the door. And on the bed lay Martin; his eyes were still open, but they were blind and unseeing no more, and he turned them wearily to the door when Helen and his father entered. But when he saw them they brightened a little. The doctor had stood back from the bed; Nurse Baker was by him. Then Martin spoke.

"It is nice to be in my own room again," he said in a voice just audible. "Oh, good morning, Helen, good morning, father. I have had horrible dreams, father. I dreamed you were angry with me. How silly. You are not angry?"

Mr. Challoner came up to the bed, and knelt there, his arm resting on the blanket.

"No, dear lad," he said. "I am not; indeed I am not."

Martin shifted his position a little.

"I'm glad," he said, "because I'm so tired. Helen, I played well, really well, did I?"

"Yes, Martin; Karl Rusoff said—he said nobody ever played better."

And she was silent because she could not say any more just then.

"And what is to-day?" asked Martin at length.

"It is Easter Sunday, dear Martin," said his father.

Martin half raised his head.

"I ought to be at Mass," he said, "but I can't. It doesn't matter, does it, if one can't?"

His father came a little closer yet.

"No, dear boy," he said, "it is Mass everywhere this morning. He was crucified, and this morning He rose again. That is all the world holds, and the heaven of heavens."

"Yes, all," said the boy. "And to-day——"

The whisper in which he had spoken died, and

Dr. Thaxter took a step towards the bed, looked at him a moment, and then went back again.

For a minute or two Martin lay there quite still: then he put out his two hands on each side of the bed, one towards Helen, one to his father.

"I am awfully tired," he said, "and I can't talk. But I can listen still. Is Stella here?"

"No, Martin," said Helen, "but she is coming as quickly as she can."

"Ah. Father, say something, something that you and I both know and like."

Mr. Challoner gently pressed the boy's hand: then he raised his head and spoke:

"The King of Love my Shepherd is,
  Whose goodness faileth never:
I nothing lack if I am His,
  And He is mine for ever."

Helen was on the other side of the bed, and as her father's voice faltered and stopped, she looked up.

"Shall father and I say it together, Martin?" she asked.

"Yes, together," said he.

So sometimes one, sometimes the other, sometimes both repeated the beautiful words. But just before the last verse Martin raised his head a little, looking straight in front of him. Then his father began:

"And so through all the length of days
  Thy goodness faileth never——"

He paused, for he saw that look in dying eyes, those eyes that were so dear to him, which means that the great event is there, that the great white presence has entered. Helen had seen too.

Then Martin raised himself a little further, and spoke no longer in a whisper:

> "Good Shepherd, may I sing Thy praise
> Within thy courts for ever."

He sank down again, withdrew his hand from his father's, and put it on the pillow. Then he laid his face on it, as was his custom, and fell asleep.

THE END.

PRINTING OFFICE OF THE PUBLISHER.

TAUCHNITZ EDITION.

February 1, 1912.

(This list appears on the first of each month, and contains all publications up to date.)

Contents:

Important Notice.

The Tauchnitz Edition in two new and attractive bindings.

The publishers beg to call attention to the fact that from now on copies of the Tauchnitz Edition can also be obtained in two entirely new and original bindings, the one in red cloth (Original-Leinenband), the other in an elegant olive-brown soft material (Original-Geschenkband). The price of each volume is ℳ 1,60 or francs 2,00 sewed, ℳ 2,20 or francs 2,75 in cloth (Original-Leinenband), and ℳ 3,00 or francs 3,75 bound in soft material (Original-Geschenkband).

A larger and more fully detailed Catalogue

of the Tauchnitz publications—containing much important and useful information in regard to authors and their works, and specially arranged for ready reference—is to be had on application of all booksellers gratis.

Particulars of the Latest Tauchnitz Volumes

will be found: firstly, within and outside the *Covers* of each volume in the Collection; secondly, on the *Bookmark* supplied with each work; and thirdly, in a *Monthly Descriptive List* supplied gratis by all booksellers.

Complete List
of the
Collection of British and American Authors.

Published with Continental Copyright by special agreement with the authors. Vide p. 1.

4312 Volumes. 417 British, 67 American Authors.
3901 Volumes by British, 411 Volumes by American Authors.

**Price of each Volume M. 1,60 or Fr. 2,00 sewed,
M. 2,20 or Fr. 2,75 in red cloth (Leinenband),
M. 3,00 or Fr. 3,75 in soft binding (Geschenkband).**

Adams, Henry (Am.): *vide* "Democracy," Author of.

Adams, Rev. W., † 1848.
Sacred Allegories 1 v.

Aguilar, Grace, † 1847.
Home Influence 2 v. — The Mother's Recompense 2 v.

Aïdé, [Charles] Hamilton, † 1906.
Rita 1 v. — Carr of Carrlyon 2 v. — The Marstons 2 v. — In that State of Life 1 v. — Morals and Mysteries 1 v. — Penruddocke 2 v. — "A nine Days' Wonder" 1 v. — Poet and Peer 2 v. — Introduced to Society 1 v.

Ainsworth, W. Harrison, † 1882.
Windsor Castle 1 v. — Saint James's 1 v. — Jack Sheppard (with Portrait) 1 v. — The Lancashire Witches 2 v. — The Star-Chamber 2 v. — The Flitch of Bacon 1 v. — The Spendthrift 1 v. — Mervyn Clitheroe 2 v. — Ovingdean Grange 1 v. — The Constable of the Tower 1 v. — The Lord Mayor of London 2 v. — Cardinal Pole 2 v. — John Law 2 v. — The Spanish Match 2 v. — The Constable de Bourbon 2 v. — Old Court 2 v. — Myddleton Pomfret 2 v. — The South-Sea Bubble 2 v. — Hilary St. Ives 2 v. — Talbot Harland 1 v. — Tower Hill 1 v. — Boscobel 2 v. — The Good Old Times 2 v. — Merry England 2 v. — The Goldsmith's Wife 2 v. — Preston Fight 2 v. — Chetwynd Calverley 2 v. — The Leaguer of Lathom 2 v. — The Fall of Somerset 2 v. — Beatrice Tyldesley 2 v. — Beau Nash 2 v. — Stanley Brereton 2 v.

Albanesi, Madame.
Drusilla's Point of View 1 v.

Alcott, Louisa M. (Am.), † 1888.
Little Women 2 v. — Little Men 1 v. — An Old-Fashioned Girl 1 v. — Jo's Boys 1 v.

Aldrich, Thomas Bailey (Am.), † 1907.
Marjorie Daw and other Tales 1 v. — The Stillwater Tragedy 1 v.

Alexander, Mrs. (Hector), † 1902.
A Second Life 3 v. — By Woman's Wit 1 v. — Mona's Choice 2 v. — A Life Interest 2 v. — A Crooked Path 2 v. — Blind Fate 2 v. — A Woman's Heart 2 v. — For His Sake 2 v. — The Snare of the Fowler 2 v. — Found Wanting 2 v. — A Ward in Chancery 1 v. — A Choice of Evils 2 v. — A Fight with Fate 2 v. — A Winning Hazard 1 v. — A Golden Autumn 1 v. — Mrs. Crichton's Creditor 1 v. — Barbara, Lady's Maid and Peeress 1 v. — The Cost of Her Pride 2 v. — Brown, V.C. 1 v. — Through Fire to Fortune 1 v. — A Missing Hero 1 v. — The Yellow Fiend 1 v. — Stronger than Love 2 v. — Kitty Costello 1 v.

Alice, Grand-Duchess of Hesse, † 1878.
Letters to Her Majesty the Queen (with Portrait). With a Memoir by H. R. H. Princess Christian 2 v.

Alldridge, Lizzie.
By Love and Law 2 v. — The World she awoke in 2 v.

Allen, Grant, † 1899.
The Woman who did 1 v.

"All for Greed," Author of (Baroness de Bury).
All for Greed 1 v. — Love the Avenger 2 v.

Anstey, F. (Guthrie).
The Giant's Robe 2 v. — A Fallen Idol 1 v. — The Pariah 3 v. — The Talking Horse and other Tales 1 v. — Voces Populi *(First and Second Series)* 1 v. — The Brass Bottle 1 v. — A Bayard from Bengal 1 v. — Salted Almonds 1 v.

Arnold, Sir Edwin, † 1904.
The Light of Asia (with Portrait) 1 v.

Arnold, Matthew, † 1888.
Essays in Criticism 2 v. — Essays in Criticism *(Second Series)* 1 v.

Atherton, Gertrude Franklin (Am.).
American Wives and English Husbands 1 v. — The Californians 1 v. — Patience Sparhawk and her Times 2 v. — Senator North 2 v. — The Doomswoman 1 v. — The Aristocrats 1 v. — The Splendid Idle Forties 1 v. — The Conqueror 2 v. — A Daughter of the Vine 1 v. — His Fortunate Grace, etc. 1 v. — The Valiant Runaways 1 v. — The Bell in the Fog, and Other Stories 1 v. — The Travelling Thirds (in Spain) 1 v. — Rezánov 1 v. — Ancestors 2 v. — The Gorgeous Isle 1 v. — Tower of Ivory 2 v.

Austen, Jane, † 1817.
Sense and Sensibility 1 v. — Mansfield Park 1 v. — Pride and Prejudice 1 v. — Northanger Abbey, and Persuasion 1 v. — Emma 1 v.

"Autobiography of Lutfullah," Author of: *vide* **E. B. Eastwick.**

Avebury, Lord: *vide* **Sir John Lubbock.**

Bagot, Richard.
A Roman Mystery 2 v. — Casting of Nets 2 v. — The Just and the Unjust 2 v. — Donna Diana 2 v. — Love's Proxy 1 v. — The Passport 2 v. — Temptation 2 v. — The Lakes of Northern Italy 1 v. — Anthony Cuthbert 2 v. — The House of Serravalle 2 v.

Baring-Gould, S.
Mehalah 1 v. — John Herring 2 v. — Court Royal 2 v.

Barker, Lady: *v.* **Lady Broome.**

Barrett, Frank.
The Smuggler's Secret 1 v. — Out of the Jaws of Death 2 v.

Barrie, J. M.
Sentimental Tommy 2 v. — Margaret Ogilvy 1 v. — Tommy and Grizel 2 v. — The Little White Bird 1 v. — Peter and Wendy 1 v.

"Bayle's Romance, Miss," Author of: *vide* **W. Fraser Rae.**

Baynes, Rev. Robert H.
Lyra Anglicana, Hymns and Sacred Songs 1 v.

Beaconsfield: *vide* **Disraeli.**

Beaumont, Averil (Mrs. Hunt).
Thornicroft's Model 2 v.

Max Beerbohm.
Zuleika Dobson 1 v.

Bell, Currer (Charlotte Brontë—Mrs. Nicholls), † 1855.
Jane Eyre 2 v. — Shirley 2 v. — Villette 2 v. — The Professor 1 v.

Bell, Ellis & Acton (Emily, † 1848, and Anne, † 1849, Brontë).
Wuthering Heights, and Agnes Grey 2 v.

Bellamy, Edward (Am.), † 1898.
Looking Backward 1 v.

Benedict, Frank Lee (Am.).
St. Simon's Niece 2 v.

Bennett, Arnold.
The Grand Babylon Hotel 1 v. — The Gates of Wrath 1 v. — A Great Man 1 v. — Sacred and Profane Love 1 v. — Whom God hath joined 1 v. — The Ghost 1 v. — The Grim Smile of the Five Towns 1 v. — Buried Alive 1 v. — The Old Wives' Tale 2 v. — The Glimpse 1 v. — Helen with the High Hand 1 v. — Clayhanger 2 v. — The Card 1 v. — Hilda Lessways 1 v.
(*Vide* **Eden Phillpotts.**)

Benson, E. F.
Dodo 1 v. — The Rubicon 1 v. — Scarlet and Hyssop 1 v. — The Book of Months 1 v. — The Relentless City 1 v. — Mammon & Co. 2 v. — The Challoners 1 v. — An Act in a Backwater 1 v. — The Image in the Sand 2 v. — The Angel of Pain 2 v. — Paul 2 v. — The House of Defence 2 v. — Sheaves 2 v. — The Climber 2 v. — The Blotting Book 1 v. — A Reaping 1 v. — Daisy's Aunt 1 v. — The Osbornes 1 v. — Account Rendered 1 v. — Juggernaut 1 v.

Benson, Robert Hugh.
The Necromancers 1 v. — A Winnowing 1 v. — None Other Gods 1 v. — The Dawn of All 1 v.

Besant, Sir Walter, † 1901.
The Revolt of Man 1 v. — Dorothy Forster 2 v. — Children of Gibeon 2 v. —

The World went very well then 2 v. — Katharine Regina 1 v. — Herr Paulus 2 v. — The Inner House 1 v. — The Bell of St. Paul's 2 v. — For Faith and Freedom 2 v. — Armorel of Lyonesse 2 v. — Verbena Camellia Stephanotis, etc. 1 v. — Beyond the Dreams of Avarice 2 v. — The Master Craftsman 2 v. — A Fountain Sealed 1 v. — The Orange Girl 2 v. — The Fourth Generation 1 v. — The Lady of Lynn 2 v.

Besant, Sir Walter, † 1901, & James Rice, † 1882.

The Golden Butterfly 2 v. — Ready-Money Mortiboy 2 v. — By Celia's Arbour 2 v.

Betham-Edwards, M.

The Sylvestres 1 v. — Felicia 2 v. — Brother Gabriel 2 v. — Forestalled 1 v. — Exchange no Robbery, and other Novelettes 1 v. — Disarmed 1 v. — Doctor Jacob 1 v. — Pearla 1 v. — Next of Kin Wanted 1 v. — The Parting of the Ways 1 v. — For One and the World 1 v. — The Romance of a French Parsonage 1 v. — France of To-day 1 v. — Two Aunts and a Nephew 1 v. — A Dream of Millions 1 v. — The Curb of Honour 1 v. — France of To-day (*Second Series*) 1 v. — A Romance of Dijon 1 v. — The Dream-Charlotte 1 v. — A Storm-Rent Sky 1 v. — Reminiscences 1 v. — The Lord of the Harvest 1 v. — Anglo-French Reminiscences, 1875—1899 1 v. — A Suffolk Courtship 1 v. — Mock Beggars' Hall 1 v. — East of Paris 1 v. — A Humble Lover 1 v. — Barham Brocklebank, M.D. 1 v. — Martha Rose, Teacher 1 v. — The White House by the Sea 1 v.

Bierce, Ambrose (Am.).

In the Midst of Life 1 v.

Birchenough, Mabel C.

Potsherds 1 v.

Bisland, E. (Am.): *vide* Rhoda Broughton.

Bismarck, Prince: *vide* Butler. *Vide* also Wilhelm Görlach (Collection of German Authors, p. 29), and Whitman.

Black, William, † 1898.

A Daughter of Heth 2 v. — In Silk Attire 2 v. — The Strange Adventures of a Phaeton 2 v. — A Princess of Thule 2 v. — Kilmeny 1 v. — The Maid of Killeena, and other Stories 1 v. — Three Feathers 2 v. — Lady Silverdale's Sweetheart, and other Stories 1 v. — Madcap Violet 2 v. — Green Pastures and Piccadilly 2 v. — Macleod of Dare 2 v. — White Wings 2 v. — Sunrise 2 v. — The Beautiful Wretch 1 v. — Mr. Pisistratus Brown, M.P., in the Highlands; The Four Macnicols; The Pupil of Aurelius 1 v. — Shandon Bells (with Portrait) 2 v. — Judith Shakespeare 2 v. — The Wise Women of Inverness, etc. 1 v. — White Heather 2 v. — Sabina Zembra 2 v. — The Strange Adventures of a House-Boat 2 v. — In Far Lochaber 2 v. — The New Prince Fortunatus 2 v. — Stand Fast, Craig-Royston! 2 v. — Donald Ross of Heimra 2 v. — The Magic Ink, and other Tales 1 v. — Wolfenberg 2 v. — The Handsome Humes 2 v. — Highland Cousins 2 v. — Briseis 2 v. — Wild Eelin 2 v.

"Black-Box Murder, the," Author of.

The Black-Box Murder 1 v.

Blackmore, Richard Doddridge, † 1900.

Alice Lorraine 2 v. — Mary Anerley 3 v. — Christowell 2 v. — Tommy Upmore 2 v. — Perlycross 2 v.

"Blackwood."

Tales from "Blackwood" (*First Series*) 1 v. — Tales from "Blackwood" (*Second Series*) 1 v.

Blagden, Isa, † 1873.

The Woman I loved, and the Woman who loved me; A Tuscan Wedding 1 v.

Blessington, Countess of (Marguerite Gardiner), † 1849.

Meredith 1 v. — Strathern 2 v. — Memoirs of a Femme de Chambre 1 v. — Marmaduke Herbert 2 v. — Country Quarters (with Portrait) 2 v.

Bloomfield, Baroness.

Reminiscences of Court and Diplomatic Life (with the Portrait of Her Majesty the Queen) 2 v.

Boldrewood, Rolf.

Robbery under Arms 2 v. — Nevermore 2 v.

Braddon, Miss (Mrs. Maxwell).

Lady Audley's Secret 2 v. — Aurora Floyd 2 v. — Eleanor's Victory 2 v. — John Marchmont's Legacy 2 v. — Henry Dunbar 2 v. — The Doctor's Wife 2 v. — Only a Clod 2 v. — Sir Jasper's Tenant 2 v. — The Lady's Mile 2 v. — Rupert Godwin 2 v. — Dead-Sea Fruit 2 v. — Run to Earth 2 v. — Fenton's Quest 2 v. — The Lovels of Arden 2 v. — Strangers and Pilgrims 2 v. — Lucius Davoren 3 v. — Taken at the Flood 3 v. — Lost for Love 2 v. — A Strange World 2 v. — Hostages to Fortune 2 v. — Dead Men's Shoes 2 v. — Joshua Haggard's Daughter 2 v. — Weavers and Weft 1 v. — In Great Waters, and other Tales 1 v. — An Open Verdict 3 v. — Vixen 3 v. — The Cloven Foot 3 v. — The Story of Barbara 2 v. — Just as I am 2 v. — Asphodel 3 v. — Mount Royal 2 v. — The Golden Calf 2 v. — Flower and Weed 1 v. — Phantom Fortune 3 v. — Under the Red Flag 1 v. — Ishmael 3 v. — Wyllard's Weird 3 v. — One Thing Needful 2 v. — Cut by the County 1 v. — Like and Unlike 2 v. — The Fatal Three 2 v. — The Day will come 2 v. — One Life, One Love 2 v. — Gerard 2 v. — The Venetians 2 v. — All along the River 2 v. — Thou art the Man 2 v. — The Christmas Hirelings, etc. 1 v. — Sons of Fire 2 v. — London Pride 2 v. — Rough Justice 2 v. — In High Places 2 v. — His Darling Sin 1 v. — The Infidel 2 v. — The Conflict 2 v. — The Rose of Life 2 v. — Dead Love has Chains 1 v. — During Her Majesty's Pleasure 1 v.

Brassey, Lady, † 1887.

A Voyage in the "Sunbeam" 2 v. — Sunshine and Storm in the East 2 v. — In the Trades, the Tropics and the Roaring Forties 2 v.

"Bread-Winners, the," Author of (Am.).

The Bread-Winners 1 v.

Bret Harte: *vide* Harte.

Brock, Rev. William, † 1875.

Sir Henry Havelock, K. C. B. 1 v.

Brontë, Charlotte: *vide* Currer Bell.

Brontë, Emily & Anne: *vide* Ellis & Acton Bell.

Brooks, Shirley, † 1874.

The Silver Cord 3 v. — Sooner or Later 3 v.

Broome, Lady (Lady Barker).

Station Life in New Zealand 1 v. — Station Amusements in New Zealand 1 v. — A Year's Housekeeping in South Africa 1 v. — Letters to Guy, and A Distant Shore—Rodrigues 1 v. — Colonial Memories 1 v.

Broughton, Rhoda.

Cometh up as a Flower 1 v. — Not wisely, but too well 2 v. — Red as a Rose is She 2 v. — Tales for Christmas Eve 1 v. — Nancy 2 v. — Joan 2 v. — Second Thoughts 2 v. — Belinda 2 v. — Doctor Cupid 2 v. — Alas! 2 v. — Mrs. Bligh 1 v. — A Beginner 1 v. — Scylla or Charybdis? 1 v. — Dear Faustina 1 v. — The Game and the Candle 1 v. — Foes in Law 1 v. — Lavinia 1 v. — Mamma 1 v. — The Devil and the Deep Sea 1 v.

Broughton, Rhoda, & Elizabeth Bisland (Am.).

A Widower Indeed 1 v.

Brown, John, † 1882.

Rab and his Friends, and other Papers 1 v.

Browning, Elizabeth Barrett, † 1861.

A Selection from her Poetry (with Portrait) 1 v. — Aurora Leigh 1 v.

Browning, Robert, † 1889.

Poetical Works (with Portrait) 4 v.

Bullen, Frank T.

The Cruise of the "Cachalot" 2 v.

Bulwer, Edward, Lord Lytton, † 1873.

Pelham (with Portrait) 1 v. — Eugene Aram 1 v. — Paul Clifford 1 v. — Zanoni 1 v. — The Last Days of Pompeii 1 v. — The Disowned 1 v. — Ernest Maltravers 1 v. — Alice 1 v. — Eva, and The Pilgrims of the Rhine 1 v. — Devereux 1 v. — Godolphin and Falkland 1 v. — Rienzi 1 v. — Night and Morning 1 v. — The Last of the Barons 2 v. — Athens 2 v. — The Poems and Ballads of Schiller 1 v. — Lucretia 2 v. — Harold 2 v. — King Arthur 2 v. — The New Timon, and St. Stephen's 1 v. — The Caxtons 2 v. — My Novel 4 v. — What will he do with it? 4 v. — Dramatic Works 2 v. — A Strange Story 2 v. — Caxtoniana 2 v. — The Lost Tales of Miletus 1 v. — Miscellaneous Prose Works 4 v. — Odes and Epodes of Horace 2 v. — Kenelm Chillingly 4 v. — The Coming Race 1 v. — The Parisians 4 v. — Pausanias, the Spartan 1 v.

Bulwer, Henry Lytton (Lord Dalling), † 1872.
Historical Characters 2 v. — The Life of Viscount Palmerston 3 v.

Bunyan, John, † 1688.
The Pilgrim's Progress 1 v.

"Buried Alone," Author of (Charles Wood).
Buried Alone 1 v.

Burnett, Mrs. Frances Hodgson (Am.).
Through one Administration 2 v. — Little Lord Fauntleroy 1 v. — Sara Crewe, and Editha's Burglar 1 v. — The Pretty Sister of José 1 v. — A Lady of Quality 2 v. — His Grace of Osmonde 2 v. — The Shuttle 2 v.

Burney, Miss (Madame D'Arblay), † 1840.
Evelina 1 v.

Burns, Robert, † 1796.
Poetical Works (with Portrait) 1 v.

Burton, Richard F., † 1890.
A Pilgrimage to Mecca and Medina 3 v.

Bury, Baroness de: *vide* "All for Greed."

Butler, A. J.
Bismarck. His Reflections and Reminiscences. Translated from the great German edition, under the supervision of A. J. Butler. With two Portraits. 3 v.

Buxton, Mrs. B. H., † 1881.
Jennie of "The Prince's," 2 v. — Won 2 v. — Great Grenfell Gardens 2 v. — Nell—on and off the Stage 2 v. — From the Wings 2 v.

Byron, Lord, † 1824.
Poetical Works (with Portrait) 5 v.

Caffyn, Mrs. Mannington (Iota).
A Yellow Aster 1 v. — Children of Circumstance 2 v. — Anne Mauleverer 2 v.

Caine, Hall.
The Bondman 2 v. — The Manxman 2 v. — The Christian 2 v. — The Eternal City 3 v. — The Prodigal Son 2 v. — The White Prophet 2 v.

Cameron, Verney Lovett.
Across Africa 2 v.

Campbell Praed: *vide* Praed.

Carey, Rosa Nouchette, † 1909.
Not Like other Girls 2 v. — "But Men must Work" 1 v. — Sir Godfrey's Granddaughters 2 v. — The Old, Old Story 2 v. — Herb of Grace 2 v. — The Highway of Fate 2 v. — A Passage Perilous 2 v. — At the Moorings 2 v.

Carlyle, Thomas, † 1881.
The French Revolution 3 v. — Frederick the Great 13 v. — Oliver Cromwell's Letters and Speeches 4 v. — The Life of Schiller 1 v.

Carnegie, Andrew (Am.).
Problems of To-Day 1 v.

Carr, Alaric.
Treherne's Temptation 2 v.

Castle, Agnes & Egerton.
The Star Dreamer 2 v. — Incomparable Bellairs 1 v. — Rose of the World 1 v. — French Nan 1 v. — "If Youth but knew!" 1 v. — My Merry Rockhurst 1 v. — Flower o' the Orange 1 v. — Wroth 2 v. — Diamond Cut Paste 1 v. — The Lost Iphigenia 1 v.

Castle, Egerton.
Consequences 2 v. — "La Bella," and Others 1 v.

Charles, Mrs. Elizabeth Rundle, † 1896: *vide* "Chronicles of the Schönberg-Cotta Family."

Charlesworth, Maria Louisa, † 1880.
Oliver of the Mill 1 v.

Chesterton, G. K.
The Man who was Thursday 1 v. — What's Wrong with the World 1 v. — The Innocence of Father Brown 1 v.

Cholmondeley, Mary.
Diana Tempest 2 v. — Red Pottage 2 v. — Moth and Rust 1 v. — Prisoners 2 v. — The Lowest Rung 1 v.

Christian, Princess: *vide* Alice, Grand Duchess of Hesse.

"Chronicles of the Schönberg-Cotta Family," Author of (Mrs. E. Rundle Charles), † 1896.
Chronicles of the Schönberg-Cotta Family 2 v. — The Draytons and the Davenants 2 v. — On Both Sides of the Sea 2 v. — Winifred Bertram 1 v. — Diary of Mrs. Kitty Trevylyan 1 v. —

The Victory of the Vanquished 1 v. — The Cottage by the Cathedral and other Parables 1 v. — Against the Stream 2 v. — The Bertram Family 2 v. — Conquering and to Conquer 1 v. — Lapsed, but not Lost 1 v.

Churchill, Winston (Am.).

Mr. Crewe's Career 2 v.

Clark, Alfred.

The Finding of Lot's Wife 1 v.

Clemens, Samuel L.: *v.* Twain.

Clifford, Mrs. W. K.

Love-Letters of a Worldly Woman 1 v. —Aunt Anne 2 v.—The Last Touches, and other Stories 1 v. — Mrs. Keith's Crime 1 v. — A Wild Proxy 1 v. — A Flash of Summer 1 v. — A Woman Alone 1 v. — Woodside Farm 1 v. — The Modern Way 1 v. — The Getting Well of Dorothy 1 v. — Mere Stories 1 v.

Clive, Mrs. Caroline, † 1873: *vide* Author of "Paul Ferroll."

Cobbe, Frances Power, † 1904.

Re-Echoes 1 v.

Coleridge, C. R.

An English Squire 2 v.

Coleridge, M. E.

The King with two Faces 2 v.

Coleridge, Samuel Taylor, † 1834.

Poems 1 v.

Collins, Charles Allston, † 1873.

A Cruise upon Wheels 2 v.

Collins, Mortimer, † 1876.

Sweet and Twenty 2 v. — A Fight with Fortune 2 v.

Collins, Wilkie, † 1889.

After Dark 1 v. — Hide and Seek 2 v. — A Plot in Private Life, etc. 1 v. — The Woman in White 2 v. — Basil 1 v. — No Name 3 v. — The Dead Secret, and other Tales 2 v. — Antonina 2 v. — Armadale 3 v. — The Moonstone 2 v. — Man and Wife 3 v. — Poor Miss Finch 2 v. — Miss or Mrs.? 1 v. — The New Magdalen 2 v. — The Frozen Deep 1 v. — The Law and the Lady 2 v. — The Two Destinies 1 v. — My Lady's Money, and Percy and the Prophet 1 v. — The Haunted Hotel 1 v. — The Fallen Leaves 2 v. — Jezebel's Daughter 2 v. — The Black Robe 2 v. — Heart and Science 2 v. — "I say No," 2 v. — The Evil Genius 2 v. — The Guilty River, and The Ghost's Touch 1 v. — The Legacy of Cain 2 v.—Blind Love 2 v.

"Cometh up as a Flower": *vide* Rhoda Broughton.

Conrad, Joseph.

An Outcast of the Islands 2 v. — Tales of Unrest 1 v. — The Secret Agent 1 v. — A Set of Six 1 v. — Under Western Eyes 1 v.

Conway, Hugh (F. J. Fargus), † 1885.

Called Back 1 v. — Bound Together 2 v. — Dark Days 1 v. — A Family Affair 2 v. — Living or Dead 2 v.

Cooper, James Fenimore (Am.), † 1851.

The Spy (with Portrait) 1 v. — The Two Admirals 1 v. — The Jack O'Lantern 1 v.

Cooper, Mrs.: *vide* Katharine Saunders.

Corelli, Marie.

Vendetta! 2 v. — Thelma 2 v. — A Romance of Two Worlds 2 v. — "Ardath" 3 v. — Wormwood. A Drama of Paris 2 v. — The Hired Baby, with other Stories and Social Sketches 1 v. — Barabbas; A Dream of the World's Tragedy 2 v. — The Sorrows of Satan 2 v. — The Mighty Atom 1 v. — The Murder of Delicia 1 v. — Ziska 1 v. — Boy. A Sketch. 2 v. — The Master-Christian 2 v.—"Temporal Power" 2 v. — God's Good Man 2 v. — Free Opinions 1 v. — Treasure of Heaven (with Portrait) 2 v. — Holy Orders 2 v. — The Life Everlasting 2 v.

Cotes, Mrs. Everard.

Those Delightful Americans 1 v. — Set in Authority 1 v. — Cousin Cinderella 1 v.

"County, the," Author of.

The County 1 v.

Craik, George Lillie, † 1866.

A Manual of English Literature and of the History of the English Language 2 v.

Craik, Mrs. (Miss Dinah M. Mulock), † 1887.

John Halifax, Gentleman 2 v. — The Head of the Family 2 v. — A Life for a Life 2 v. — A Woman's Thoughts about Women 1 v. — Agatha's Husband 1 v. — Romantic Tales 1 v. — Domestic Stories 1 v. — Mistress and Maid 1 v. — The Ogilvies 1 v. — Lord Erlistoun 1 v. — Christian's Mistake 1 v. — Bread upon the Waters 1 v. — A Noble Life 1 v. — Olive 2 v. — Two Marriages 1 v. — Studies from Life 1 v. — Poems 1 v. — The Woman's Kingdom 2 v. — The Unkind Word, and other Stories 2 v. — A Brave Lady 2 v. — Hannah 2 v. — Fair France 1 v. — My Mother and I 1 v. — The Little Lame Prince 1 v. — Sermons out of Church 1 v.—The Laurel-Bush; Two little Tinkers 1 v. —A Legacy 2 v.— Young Mrs. Jardine 2 v. — His Little Mother, and other Tales and Sketches 1 v. — Plain Speaking 1 v. — Miss Tommy 1 v. — King Arthur 1 v.

Craik, Georgiana M. (Mrs. May).

Lost and Won 1 v. — Faith Unwin's Ordeal 1 v. — Leslie Tyrrell 1 v. —Winifred's Wooing, etc. 1 v. — Mildred 1 v. — Esther Hill's Secret 2 v. — Hero Trevelyan 1 v. — Without Kith or Kin 2 v. — Only a Butterfly 1 v. — Sylvia's Choice; Theresa 2 v. — Anne Warwick 1 v. — Dorcas 2 v. — Two Women 2 v.

Craik, Georgiana M., & M. C. Stirling.

Two Tales of Married Life (Hard to Bear, by Miss Craik: A True Man, by M. C. Stirling) 2 v.

Craven, Mrs. Augustus: *vide* Lady Fullerton.

Crawford, F. Marion (Am.), † 1909.

Mr. Isaacs 1 v. — Doctor Claudius 1 v. — To Leeward 1 v. — A Roman Singer 1 v. — An American Politician 1 v. — Zoroaster 1 v. —A Tale of a Lonely Parish 2 v. — Saracinesca 2 v. — Marzio's Crucifix 1 v.—Paul Patoff 2 v.—With the Immortals 1 v. — Greifenstein 2 v. — Sant' Ilario 2 v. — A Cigarette-Maker's Romance 1 v. — Khaled 1 v. — The Witch of Prague 2 v. — The Three Fates 2 v. — Don Orsino 2 v. — The Children of the King 1 v. — Pietro Ghisleri 2 v. — Marion Darche 1 v. — Katharine Lauderdale 2 v. — The Ralstons 2 v. — Casa Braccio 2 v. — Adam Johnstone's Son 1 v. — Taquisara 2 v. — A Rose of Yesterday 1 v. — Corleone 2 v. — Via Crucis 2 v. — In the Palace of the King 2 v. — Marietta, a Maid of Venice 2 v. — Cecilia 2 v. — The Heart of Rome 2 v. — Whosoever Shall Offend... 2 v. — Soprano 2 v. — A Lady of Rome 2 v. — Arethusa 2 v. — The Primadonna 2 v.— The Diva's Ruby 2 v. — The White Sister 1 v. — Stradella 1 v. — The Undesirable Governess 1 v. — Uncanny Tales 1 v.

Crockett, S. R.

The Raiders 2 v. — Cleg Kelly 2 v.— The Grey Man 2 v. — Love Idylls 1 v. — The Dark o' the Moon 2 v.

Croker, B. M.

Peggy of the Bartons 2 v. — The Happy Valley 1 v. — The Old Cantonment, with Other Stories of India and Elsewhere 1 v. — A Nine Days' Wonder 1 v. — The Youngest Miss Mowbray 1 v. — The Company's Servant 2 v. — The Cat's-Paw 1 v. — Katherine the Arrogant 1 v. — Fame 1 v. — Babes in the Wood 1 v. — A Rolling Stone 1 v.

Cross, J. W.: *vide* George Eliot's Life.

Cudlip, Mrs. Pender: *vide* A. Thomas.

Cummins, Miss (Am.), † 1866.

The Lamplighter 1 v. — Mabel Vaughan 1 v.—El Fureidîs 1 v.—Haunted Hearts 1 v.

Cushing, Paul.

The Blacksmith of Voe 2 v.

"Daily News."

War Correspondence, 1877, by Archibald Forbes and others 3 v.

Danby, Frank.

The Heart of a Child 2 v. — An Incompleat Etonian 2 v. — Let the Roof fall in 2 v.

"Dark," Author of.

Dark 1 v.

Davis, Richard Harding (Am.).

Gallegher, etc. 1 v. — Van Bibber and Others 1 v. — Ranson's Folly 1 v.

De Foe, Daniel, † 1731.

Robinson Crusoe 1 v.

Deland, Margaret (Am.).

John Ward, Preacher 1 v.

"Democracy," Author of (Am.).

Democracy 1 v.

De Morgan, William.

Joseph Vance 2 v.

"Demos," Author of: *vide* George Gissing.

De Quincey, Thomas.

Confessions of an English Opium-Eater 1 v.

"Diary and Notes": *vide* Author of "Horace Templeton."

Dickens, Charles, † 1870.

The Pickwick Club (with Portrait) 2 v. — American Notes 1 v. — Oliver Twist 1 v. — Nicholas Nickleby 2 v. — Sketches 1 v. — Martin Chuzzlewit 2 v. — A Christmas

Carol; The Chimes; The Cricket on the Hearth 1 v. — Master Humphrey's Clock (Old Curiosity Shop; Barnaby Rudge, etc.) 3 v. — Pictures from Italy 1 v. — Dombey and Son 3 v. — David Copperfield 3 v. — Bleak House 4 v. — A Child's History of England (2 v. 8⁰ M. 2,70.) — Hard Times 1 v. — Little Dorrit (with Illustrations) 4 v. — The Battle of Life; The Haunted Man 1 v. — A Tale of two Cities 2 v. — Hunted Down; The Uncommercial Traveller 1 v. — Great Expectations 2 v. — Christmas Stories, etc. 1 v. — Our Mutual Friend (with Illustrations) 4 v. — Somebody's Luggage; Mrs. Lirriper's Lodgings; Mrs. Lirriper's Legacy 1 v. — Doctor Marigold's Prescriptions; Mugby Junction 1 v. — The Mystery of Edwin Drood (with Illustrations) 2 v. — The Mudfog Papers, 1 v. — The Letters of Charles Dickens, ed. by his Sister-in-law and his eldest Daughter 4 v. — *Vide* also Household Words, Novels and Tales, and John Forster.

Dickens, Charles, & Wilkie Collins.

No Thoroughfare; The Late Miss Hollingford 1 v.

Disraeli, Benjamin, Lord Beaconsfield, † 1881.

Coningsby 1 v. — Sybil 1 v. — Contarini Fleming (with Portrait) 1 v. — Alroy 1 v. — Tancred 2 v. — Venetia 2 v. — Vivian Grey 2 v. — Henrietta Temple 1 v. — Lothair 2 v. — Endymion 2 v.

Dixon, Ella Hepworth.

The Story of a Modern Woman 1 v. — One Doubtful Hour 1 v.

Dixon, W. Hepworth, † 1879.

Personal History of Lord Bacon 1 v. — The Holy Land 2 v. — New America 2 v. — Spiritual Wives 2 v. — Her Majesty's Tower 4 v. — Free Russia 2 v. — History of two Queens 6 v. — White Conquest 2 v. — Diana, Lady Lyle 2 v.

Dixon, Jr., Thomas, (Am.).

The Leopard's Spots 2 v.

Dougall, L.

Beggars All 2 v.

Dowie, Ménie Muriel.

A Girl in the Karpathians 1 v.

Doyle, Sir A. Conan.

The Sign of Four 1 v. — Micah Clarke 2 v. — The Captain of the Pole-Star, and other Tales 1 v. — The White Company 2 v. — A Study in Scarlet 1 v. — The Great Shadow, and Beyond the City 1 v. — The Adventures of Sherlock Holmes 2 v. — The Refugees 2 v. — The Firm of Girdlestone 2 v. — The Memoirs of Sherlock Holmes 2 v. — Round the Red Lamp 1 v. — The Stark Munro Letters 1 v. — The Exploits of Brigadier Gerard 1 v. — Rodney Stone 2 v. — Uncle Bernac 1 v. — The Tragedy of the Korosko 1 v. — A Duet 1 v. — The Green Flag 1 v. — The Great Boer War 2 v. — The War in South Africa 1 v. — The Hound of the Baskervilles 1 v. — Adventures of Gerard 1 v. — The Return of Sherlock Holmes 2 v. — Sir Nigel 2 v. — Through the Magic Door 1 v. — Round the Fire Stories 1 v. — The Mystery of Cloomber 1 v. — The Last Galley 1 v.

Drummond, Professor Henry, † 1897.

The Greatest Thing in the World; Pax Vobiscum; The Changed Life 1 v.

Dufferin, the Earl of.

Letters from High Latitudes 1 v.

Duncan, Sara Jeannette: *vide* Mrs. Cotes.

Dunton: *vide* Th. Watts-Dunton.

Earl, the, and the Doctor.

South Sea Bubbles 1 v.

Eastwick, Edward B., † 1883.

Autobiography of Lutfullah 1 v.

Edgeworth, Maria, *vide* Series for the Young, p. 29.

Edwardes, Mrs. Annie.

Archie Lovell 2 v. — Steven Lawrence, Yeoman 2 v. — Ought we to visit her? 2 v. — A Vagabond Heroine 1 v. — Leah: A Woman of Fashion 2 v. — A Blue-Stocking 1 v. — Jet: Her Face or Her Fortune? 1 v. — Vivian the Beauty 1 v. — A Ballroom Repentance 2 v. — A Girton Girl 2 v. — A Playwright's Daughter, and Bertie Griffiths 1 v. — Pearl-Powder 1 v. The Adventuress 1 v.

Edwards, Amelia B., † 1892.

Barbara's History 2 v. — Miss Carew 2 v. — Hand and Glove 1 v. — Half a Million of Money 2 v. — Debenham's Vow 2 v. — In the Days of my Youth 2 v. — Untrodden Peaks and Unfrequented Valleys 1 v. — Monsieur Maurice 1 v. — A Night on the Borders of the Black Forest 1 v. — A Poetry-Book of Elder Poets

1 v. — A Thousand Miles up the Nile 2 v. — A Poetry-Book of Modern Poets 1 v. — Lord Brackenbury 2 v.

Edwards, M. Betham-: *vide* Betham.

Eggleston, Edward (Am.), † 1902.
The Faith Doctor 2 v.

Elbon, Barbara (Am.).
Bethesda 2 v.

Eliot, George (Miss Evans—Mrs. Cross), † 1880.
Scenes of Clerical Life 2 v. — Adam Bede 2 v. — The Mill on the Floss 2 v. — Silas Marner 1 v. — Romola 2 v. — Felix Holt 2 v. — Daniel Deronda 4 v. — The Lifted Veil, and Brother Jacob 1 v. — Impressions of Theophrastus Such 1 v. — Essays and Leaves from a Note-Book 1 v. — George Eliot's Life, edited by her Husband, J. W. Cross 4 v.

"Elizabeth and her German Garden," Author of.
Elizabeth and her German Garden 1 v. — The Solitary Summer 1 v. — The Benefactress 2 v. — Princess Priscilla's Fortnight 1 v. — The Adventures of Elizabeth in Rügen 1 v. — Fräulein Schmidt and Mr. Anstruther 1 v.

Elliot, Mrs. Frances, † 1898.
Diary of an Idle Woman in Italy 2 v. — Old Court Life in France 2 v. — The Italians 2 v. — The Diary of an Idle Woman in Sicily 1 v. — Pictures of Old Rome 1 v. — The Diary of an Idle Woman in Spain 2 v. — The Red Cardinal 1 v. — The Story of Sophia 1 v. — Diary of an Idle Woman in Constantinople 1 v. — Old Court Life in Spain 2 v. — Roman Gossip 1 v.

Emerson, Ralph Waldo (Am.), † 1882.
Representative Men 1 v.

"Englishwoman's Love-Letters, an," Author of.
An Englishwoman's Love-Letters 1 v.

Erroll, Henry.
An Ugly Duckling 1 v.

Esler, E. Rentoul.
The Way they loved at Grimpat 1 v.

"Essays and Reviews," the Authors of.
Essays and Reviews. By various Authors 1 v.

"Estelle Russell," Author of.
Estelle Russell 2 v.

Esterre-Keeling, Elsa D'.
Three Sisters 1 v. — A Laughing Philosopher 1 v. — The Professor's Wooing 1 v. — In Thoughtland and in Dreamland 1 v. — Orchardscroft 1 v. — Appassionata 1 v. — Old Maids and Young 2 v. — The Queen's Serf 1 v.

"Euthanasia," Author of.
Euthanasia 1 v.

Ewing, Juliana Horatia, † 1885.
Jackanapes; The Story of a Short Life; Daddy Darwin's Dovecot 1 v. — A Flat Iron for a Farthing 1 v. — The Brownies, and other Tales 1 v.

"Expiated," Author of.
Expiated 2 v.

Fargus, F. J.: *vide* Hugh Conway.

Farrar, F. W. (Dean), † 1903.
Darkness and Dawn 3 v.

"Fate of Fenella, the," Authors of.
The Fate of Fenella, by 24 Authors 1 v.

Felkin, Alfred Laurence: *vide* E. T. Fowler.

Felkin, Mrs.: *vide* E. T. Fowler.

Fendall, Percy: *vide* F. C. Philips.

Fenn, George Manville.
The Parson o' Dumford 2 v. — The Clerk of Portwick 2 v.

Fielding, Henry, † 1754.
Tom Jones 2 v.

Findlater, Mary & Jane (Am.): *vide* Kate Douglas Wiggin.

Fitzgerald, Edward.
Rubáiyát of Omar Khayyám 1 v.

Five Centuries
of the English Language and Literature: John Wycliffe. — Geoffrey Chaucer. — Stephen Hawes. — Sir Thomas More. — Edmund Spenser. — Ben Jonson. — John Locke. — Thomas Gray (vol. 500, published 1860) 1 v.

Fleming, George (Am.).
Kismet 1 v. — Andromeda 2 v.

Forbes, Archibald, † 1900.
My Experiences of the War between France and Germany 2 v. — Soldiering and Scribbling 1 v. — Memories and Studies of War and Peace 2 v. — *Vide* also "Daily News," War Correspondence.

Forrest, R. E.
Eight Days 2 v.

Forrester, Mrs.
Viva 2 v. — Rhona 2 v. — Roy and Viola 2 v. — My Lord and My Lady 2 v. — I have Lived and Loved 2 v. — June 2 v. — Omnia Vanitas 1 v. — Although he was a Lord, and other Tales 1 v. — Corisande, and other Tales 1 v. — Once Again 2 v. — Of the World, Worldly 1 v. — Dearest 2 v. — The Light of other Days 1 v. — Too Late Repented 1 v.

Forster, John, † 1876.
The Life of Charles Dickens (with Illustrations and Portraits) 6 v. — Life and Times of Oliver Goldsmith 2 v.

Fothergill, Jessie.
The First Violin 2 v. — Probation 2 v. — Made or Marred, and "One of Three" 1 v. — Kith and Kin 2 v. — Peril 2 v. — Borderland 2 v.

"Found Dead," Author of: *vide* James Payn.

Fowler, Ellen Thorneycroft (Mrs. Alfred Laurence Felkin).
A Double Thread 2 v. — The Farringdons 2 v. — Fuel of Fire 1 v. — Place and Power 2 v. — In Subjection 2 v. — Miss Fallowfield's Fortune 1 v.

Fowler, Ellen Thorneycroft (Mrs. A. L. Felkin) & Alfred Laurence Felkin.
Kate of Kate Hall 2 v.

Fox, Caroline, † 1871.
Memories of Old Friends from her Journals and Letters, edited by Horace N. Pym 2 v.

"Frank Fairlegh," Author of (F. E. Smedley), † 1864.
Frank Fairlegh 2 v.

Francis, M. E.
The Duenna of a Genius 1 v.

Frederic, Harold (Am.), † 1898.
Illumination 2 v. — March Hares 1 v.

Freeman, Edward A., † 1892.
The Growth of the English Constitution 1 v. — Select Historical Essays 1 v. — Sketches from French Travel 1 v.

Froude, James Anthony, † 1894.
Oceana 1 v. — The Spanish Story of the Armada, and other Essays 1 v.

Fullerton, Lady Georgiana, † 1885.
Ellen Middleton 1 v. — Grantley Manor 2 v. — Lady Bird 2 v. — Too Strange not to be True 2 v. — Constance Sherwood 2 v. — A Stormy Life 2 v. — Mrs. Geralds' Niece 2 v. — The Notary's Daughter 1 v. — The Lilies of the Valley, and The House of Penarvan 1 v. — The Countess de Bonneval 1 v. — Rose Leblanc 1 v. — Seven Stories 1 v. — The Life of Luisa de Carvajal 1 v. — A Will and a Way, and The Handkerchief at the Window 2 v. — Eliane 2 v. (by Mrs. Augustus Craven, translated by Lady Fullerton). — Laurentia 1 v.

Galsworthy, John.
The Man of Property 2 v. — The Country House 1 v. — Fraternity 1 v. — Villa Rubein 1 v. — A Man of Devon, etc. 1 v. — A Motley 1 v. — The Patrician 1 v.

Gardiner, Marguerite: *vide* Lady Blessington.

Gaskell, Mrs., † 1865.
Mary Barton 1 v. — Ruth 2 v. — North and South 1 v. — Lizzie Leigh, and other Tales 1 v. — The Life of Charlotte Brontë 2 v. — Lois the Witch, etc. 1 v. — Sylvia's Lovers 2 v. — A Dark Night's Work 1 v. — Wives and Daughters 3 v. — Cranford 1 v. — Cousin Phillis, and other Tales 1 v.

"Geraldine Hawthorne," Author of: *v.* Author of "Miss Molly."

Gerard, Dorothea (Madame Longard de Longgarde).
Lady Baby 2 v. — Recha 1 v. — Orthodox 1 v. — The Wrong Man 1 v. — A Spotless Reputation 1 v. — A Forgotten Sin 1 v. — One Year 1 v. — The Supreme Crime 1 v. — The Blood-Tax 1 v. — Holy Matrimony 1 v. — The Eternal Woman 1 v. — Made of Money 1 v. — The Bridge of Life 1 v. — The Three Essentials 1 v. — The Improbable Idyl 1 v. — The Compromise 2 v. — Itinerant Daughters 1 v. — Restitution 1 v. — Pomp and Circumstance 1 v. — The Grass Widow 1 v. — The Inevitable Marriage 1 v. — A Glorious Lie 1 v.

Gerard, E. (Emily de Laszowska).
A Secret Mission 1 v. — A Foreigner 2 v. — The Extermination of Love 2 v.

Giberne, Agnes.
The Curate's Home 1 v.

Gissing, George, † 1903.
Demos 2 v. — New Grub Street 2 v.

Gladstone, Rt. Hon. W. E., † 1898.
Rome and the Newest Fashions in Religion 1 v. — Bulgarian Horrors, and Russia in Turkistan, with other Tracts 1 v. — The Hellenic Factor in the Eastern Problem, with other Tracts 1 v.

Glyn, Elinor.
The Visits of Elizabeth 1 v. — The Reflections of Ambrosine 1 v. — The Vicissitudes of Evangeline 1 v. — Beyond the Rocks 1 v. — Three Weeks 1 v. — Elizabeth Visits America 1 v. — His Hour 1 v. — The Reason Why 1 v.

Godfrey, Hal: *vide* **Charlotte O'Conor Eccles.**

Goldsmith, Oliver, † 1774.
Select Works (with Portrait) 1 v.

Goodman, Edward J.
Too Curious 1 v.

Gordon, Julien (Am.).
A Diplomat's Diary 1 v.

Gordon, Major-Gen. C. G., † 1885.
His Journals at Kartoum (with eighteen Illustrations) 2 v.

Gore, Mrs., † 1861.
Castles in the Air 1 v. — The Dean's Daughter 2 v. — Progress and Prejudice 2 v. — Mammon 2 v. — A Life's Lessons 2 v. — The Two Aristocracies 2 v. — Heckington 2 v.

Grand, Sarah.
Our Manifold Nature 1 v. — Babs the Impossible 2 v. — Emotional Moments 1 v.

Grant, Miss.
Victor Lescar 2 v. — The Sun-Maid 2 v. — My Heart's in the Highlands 2 v. — Artiste 2 v. — Prince Hugo 2 v. — Cara Roma 2 v.

Gray, Maxwell.
The Silence of Dean Maitland 2 v. — The Reproach of Annesley 2 v.

Grenville: Murray, E. C. (Trois-Etoiles), † 1881.
The Member for Paris 2 v. — Young Brown 2 v. — The Boudoir Cabal 3 v. — French Pictures in English Chalk *(First Series)* 2 v. — The Russians of To-day 1 v. — French Pictures in English Chalk *(Second Series)* 2 v. — Strange Tales 1 v. — That Artful Vicar 2 v. — Six Months in the Ranks 1 v. — People I have met 1 v.

Grimwood, Ethel St. Clair.
My Three Years in Manipur (with Portrait) 1 v.

Grohman, W. A. Baillie.
Tyrol and the Tyrolese 1 v.

Gunter, A. C. (Am.), † 1907.
Mr. Barnes of New York 1 v.

Guthrie, F. Anstey: *vide* **Anstey.**

"Guy Livingstone," Author of (George Alfred Laurence), † 1876.
Guy Livingstone 1 v. — Sword and Gown 1 v. — Barren Honour 1 v. — Border and Bastille 1 v. — Maurice Dering 1 v. — Sans Merci 2 v. — Breaking a Butterfly 2 v. — Anteros 2 v. — Hagarene 2 v.

Habberton, John (Am.).
Helen's Babies & Other People's Children 1 v. — The Bowsham Puzzle 1 v. — One Tramp; Mrs. Mayburn's Twins 1 v.

Haggard, H. Rider.
King Solomon's Mines 1 v. — She 2 v. — Jess 2 v. — Allan Quatermain 2 v. — The Witch's Head 2 v. — Maiwa's Revenge 1 v. — Mr. Meeson's Will 1 v. — Colonel Quaritch, V. C. 2 v. — Cleopatra 2 v. — Allan's Wife 1 v. — Beatrice 2 v. — Dawn 2 v. — Montezuma's Daughter 2 v. — The People of the Mist 2 v. — Joan Haste 2 v. — Heart of the World 2 v. — The Wizard 1 v. — Doctor Therne 1 v. — Swallow 2 v. — Black Heart and White Heart, and Elissa 1 v. — Lysbeth 2 v. — A Winter Pilgrimage 2 v. — Pearl-Maiden 2 v. — Stella Fregelius 2 v. — The Brethren 2 v. — Ayesha. The Return of 'She' 2 v. — The Way of the Spirit 2 v. — Benita 1 v. — Fair Margaret 2 v. — The Lady of Blossholme 1 v. — Morning Star 1 v. — Queen Sheba's Ring 1 v. — Red Eve 1 v.

Haggard, H. Rider, & Andrew Lang.
The World's Desire 2 v.

Hake, A. E.: *vide* **Gen. Gordon.**

Hall, Mrs. S. C., † 1881.
Can Wrong be Right? 1 v. — Marian 2 v.

Hamerton, Philip Gilbert, † 1894.
Marmorne 1 v. — French and English 2 v.

Hardy, Miss Iza: *vide* Author of "Not Easily Jealous."

Hardy, Thomas.

The Hand of Ethelberta 2 v. — Far from the Madding Crowd 2 v. — The Return of the Native 2 v. — The Trumpet-Major 2 v. — A Laodicean 2 v. — Two on a Tower 2 v. — A Pair of Blue Eyes 2 v. — A Group of Noble Dames 1 v. — Tess of the D'Urbervilles 2 v. — Life's Little Ironies 1 v. — Jude the Obscure 2 v.

Harland, Henry (Am.), † 1905.

The Cardinal's Snuff-Box 1 v. — The Lady Paramount 1 v. — My Friend Prospero 1 v. — The Royal End 1 v.

Harraden, Beatrice.

Ships that pass in the Night 1 v. — In Varying Moods 1 v. — Hilda Strafford, and The Remittance Man 1 v. — The Fowler 2 v. — Katharine Frensham 2 v. — The Scholar's Daughter 1 v. — Interplay 2 v.

Harrison, Agnes.

Martin's Vineyard 1 v.

Harrison, Mrs.: *v.* Lucas Malet.

Harte, Bret (Am.), † 1902.

Prose and Poetry (Tales of the Argonauts: — The Luck of Roaring Camp; The Outcasts of Poker Flat, etc. — Spanish and American Legends; Condensed Novels; Civic and Character Sketches; Poems) 2 v. — Idyls of the Foothills 1 v. — Gabriel Conroy 2 v. — Two Men of Sandy Bar 1 v. — Thankful Blossom, and other Tales 1 v. — The Story of a Mine 1 v. — Drift from Two Shores 1 v. — An Heiress of Red Dog, and other Sketches 1 v. — The Twins of Table Mountain, and other Tales 1 v. — Jeff Briggs's Love Story, and other Tales 1 v. — Flip, and other Stories 1 v. — On the Frontier 1 v. — By Shore and Sedge 1 v. — Maruja 1 v. — Snow-bound at Eagle's, and Devil's Ford 1 v. — The Crusade of the "Excelsior" 1 v. — A Millionaire of Rough-and-Ready, and other Tales 1 v. — Captain Jim's Friend, and the Argonauts of North Liberty 1 v. — Cressy 1 v. — The Heritage of Dedlow Marsh, and other Tales 1 v. — A Waif of the Plains 1 v. — A Ward of the Golden Gate 1 v. — A Sappho of Green Springs, and other Tales 1 v. — A First Family of Tasajara 1 v. — Colonel Starbottle's Client, and some other People 1 v. — Susy 1 v. — Sally Dows, etc. 1 v. — A Protégée of Jack Hamlin's, etc. 1 v. — The Bell-Ringer of Angel's, etc. 1 v. — Clarence 1 v. — In a Hollow of the Hills, and The Devotion of Enriquez 1 v. — The Ancestors of Peter Atherly, etc. 1 v. — Three Partners 1 v. — Tales of Trail and Town 1 v. — Stories in Light and Shadow 1 v. — Mr. Jack Hamlin's Mediation, and other Stories 1 v. — From Sand-Hill to Pine 1 v. — Under the Redwoods 1 v. — On the Old Trail 1 v. — Trent's Trust 1 v.

Hawthorne, Nathaniel (Am.), † 1864.

The Scarlet Letter 1 v. — Transformation (The Marble Faun) 2 v. — Passages from his English Note-Books 2 v.

Hay, John (Am.), † 1905: *v.* "The Bread Winners," Author of.

Hearn, Lafcadio, † 1906.

Kokoro 1 v. — Kwaidan 1 v. — Glimpses of Unfamiliar Japan *(First Series)* 1 v. — Glimpses of Unfamiliar Japan *(Second Series)* 1 v. — Gleanings in Buddha-Fields 1 v. — Out of the East 1 v. — The Romance of the Milky Way, etc. 1 v.

Hector, Mrs.: *vide* Mrs. Alexander.

"Heir of Redclyffe, the," Author of: *vide* Charlotte M. Yonge.

Helps, Sir Arthur, † 1875.

Friends in Council 2 v. — Ivan de Biron 2 v.

Hemans, Mrs. Felicia, † 1835.

Select Poetical Works 1 v.

Hewlett, Maurice.

The Forest Lovers 1 v. — Little Novels of Italy 1 v. — The Life and Death of Richard Yea-and-Nay 2 v. — New Canterbury Tales 1 v. — The Queen's Quair; or, The Six Years' Tragedy 2 v. — Fond Adventures 1 v. — The Fool Errant 2 v. — The Stooping Lady 1 v. — The Spanish Jade 1 v. — Halfway House 2 v. — Open Country 1 v. — Rest Harrow 1 v. — Brazenhead the Great 1 v. — The Song of Renny 1 v.

Hichens, Robert.

Flames 2 v. — The Slave 2 v. — Felix 2 v. — The Woman with the Fan 2 v. — The Garden of Allah 2 v. — The Black Spaniel, and Other Stories 1 v. — The Call of the Blood 2 v. — A Spirit in Prison 2 v. — Barbary Sheep 1 v. — Bella Donna 2 v. — The Spell of Egypt 1 v. — The Dweller on the Threshold 1 v. — The Fruitful Vine 2 v.

Hobart Pasha, Admiral, † 1886.

Sketches from my Life 1 v.

Hobbes, John Oliver (Mrs. Craigie) (Am.), † 1906.

The Gods, Some Mortals and Lord Wickenham 1 v. — The Serious Wooing 1 v. — The Dream and the Business 2 v.

Hoey, Mrs. Cashel.

A Golden Sorrow 2 v. — Out of Court 2 v.

Holdsworth, Annie E.

The Years that the Locust hath Eaten 1 v. — The Gods Arrive 1 v. — The Valley of the Great Shadow 1 v. — Great Lowlands 1 v. — A Garden of Spinsters 1 v.

Holme Lee: *vide* Harriet Parr.

Holmes, Oliver Wendell (Am.), † 1894.

The Autocrat of the Breakfast-Table 1 v. — The Professor at the Breakfast-Table 1 v. — The Poet at the Breakfast Table 1 v. — Over the Teacups 1 v.

Hope, Anthony (Hawkins).

Mr. Witt's Widow 1 v. — A Change of Air 1 v. — Half a Hero 1 v. — The Indiscretion of the Duchess 1 v. — The God in the Car 1 v. — The Chronicles of Count Antonio 1 v. — Comedies of Courtship 1 v. — The Heart of Princess Osra 1 v. — Phroso 2 v. — Simon Dale 2 v. — Rupert of Hentzau 1 v. — The King's Mirror 2 v. — Quisanté 1 v. — Tristram of Blent 2 v. — The Intrusions of Peggy 2 v. — Double Harness 2 v. — A Servant of the Public 2 v. — Sophy of Kravonia 2 v. — Tales of Two People 2 v. — The Great Miss Driver 2 v. — Mrs. Maxon Protests 1 v.

Hopkins, Tighe.

An Idler in Old France 1 v. — The Man in the Iron Mask 1 v. — The Dungeons of Old Paris 1 v. — The Silent Gate 1 v. — The Women Napoleon Loved 1 v.

"Horace Templeton," Author of.

Diary and Notes 1 v.

Hornung, Ernest William.

A Bride from the Bush 1 v. — Under Two Skies 1 v. — Tiny Luttrell 1 v. — The Boss of Taroomba 1 v. — My Lord Duke 1 v. — Young Blood 1 v. — Some Persons Unknown 1 v. — The Amateur Cracksman 1 v. — The Rogue's March 1 v. — The Belle of Toorak 1 v. — Peccavi 1 v. — The Black Mask 1 v. — The Shadow of the Rope 1 v. — No Hero 1 v. — Denis Dent 1 v. — Irralie's Bushranger and The Unbidden Guest 1 v. — Stingaree 1 v. — A Thief in the Night 1 v. — Dead Men Tell No Tales 1 v. — Mr. Justice Raffles 1 v. — The Camera Fiend 1 v.

"Household Words."

Conducted by Charles Dickens. 1851-56. 36 v. — NOVELS and TALES reprinted from Household Words by Charles Dickens. 1856-59. 11 v.

Houstoun, Mrs.: *vide* "Recommended to Mercy."

"How to be Happy though Married," Author of.

How to be Happy though Married 1 v.

Howard, Blanche Willis (Am.), † 1898.

One Summer 1 v. — Aunt Serena 1 v. — Guenn 2 v. — Tony, the Maid, etc. 1 v. — The Open Door 2 v.

Howard, Blanche Willis, † 1898, & William Sharp (Am.), † 1905.

A Fellowe and His Wife 1 v.

Howells, William Dean (Am.).

A Foregone Conclusion 1 v. — The Lady of the Aroostook 1 v. — A Modern Instance 2 v. — The Undiscovered Country 1 v. — Venetian Life (with Portrait) 1 v. — Italian Journeys 1 v. — A Chance Acquaintance 1 v. — Their Wedding Journey 1 v. — A Fearful Responsibility, and Tonelli's Marriage 1 v. — A Woman's Reason 2 v. — Dr. Breen's Practice 1 v. — The Rise of Silas Lapham 2 v. — A Pair of Patient Lovers 1 v. — Miss Bellard's Inspiration 1 v.

Hughes, Thomas, † 1898.

Tom Brown's School-Days 1 v.

Hungerford, Mrs. (Mrs. Argles), † 1897.

Molly Bawn 2 v. — Mrs. Geoffrey 2 v. — Faith and Unfaith 2 v. — Portia 2 v. — Loÿs, Lord Berresford, and other Tales 1 v. — Her First Appearance, and other Tales 1 v. — Phyllis 2 v. — Rossmoyne 2 v. — Doris 2 v. — A Maiden all Forlorn, etc. 1 v. — A Passive Crime, and other Stories 1 v. — Green Pleasure and Grey Grief 2 v. — A Mental Struggle 2 v. — Her Week's Amusement, and Ugly Barrington 1 v. — Lady Branksmere 2 v. — Lady Valworth's Diamonds 1 v. — A Modern Circe 2 v. — Marvel 2 v. — The Hon. Mrs. Vereker 1 v. — Under-Currents 2 v. — In Durance Vile, etc. 1 v. — A Troublesome Girl, and other Stories 1 v. — A Life's Remorse 2 v. — A Born Coquette 2 v. — The Duchess 1 v. — Lady Verner's Flight 1 v. — A Conquering Heroine, and "When in Doubt" 1 v. — Nora

Creina 2 v. — A Mad Prank, and other Stories 1 v. — The Hoyden 2 v. — The Red House Mystery 1 v. — An Unsatisfactory Lover 1 v. — Peter's Wife 2 v. — The Three Graces 1 v. — A Tug of War 1 v. — The Professor's Experiment 2 v. — A Point of Conscience 2 v. — A Lonely Girl 1 v. — Lovice 1 v. — The Coming of Chloe 1 v.

Hunt, Mrs.: *vide* Beaumont.

Hunt, Violet.

The Human Interest 1 v. — White Rose of Weary Leaf 2 v. — The Wife of Altamont 1 v. — Tales of the Uneasy 1 v.

Hutten, Baroness von (Am.).

The Halo 1 v.—Kingsmead 1 v.—The Lordship of Love 2 v. — The Green Patch 1 v.

Ingelow, Jean, † 1897.

Off the Skelligs 3 v. — Poems 2 v. — Fated to be Free 2 v. — Sarah de Berenger 2 v. — Don John 2 v.

Inglis, the Hon. Lady.

The Siege of Lucknow 1 v.

Ingram, John H.: *vide* Poe.

Iota: *vide* Mrs. Caffyn.

Irving, Washington (Am.), † 1859.

The Sketch Book (with Portrait) 1 v. — The Life of Mahomet 1 v. — Lives of the Successors of Mahomet 1 v.—Oliver Goldsmith 1 v. — Chronicles of Wolfert's Roost 1 v. — Life of George Washington 5 v.

Jackson, Mrs. Helen (H. H.) (Am.), † 1885.

Ramona 2 v.

Jacobs, W. W.

Many Cargoes 1 v. — The Skipper's Wooing, and The Brown Man's Servant 1 v. — Sea Urchins 1 v. — A Master of Craft 1 v. — Light Freights 1 v. — At Sunwich Port 1 v. — The Lady of the Barge 1 v. — Odd Craft 1 v. — Dialstone Lane 1 v. — Captains All 1 v. — Short Cruises 1 v. — Salthaven 1 v. — Sailors' Knots 1 v. — Ship's Company 1 v.

James, Charles T. C.

Holy Wedlock 1 v.

James, G. P. R., † 1860.

Morley Ernstein (with Portrait) 1 v. — Forest Days 1 v. — The False Heir 1 v. — Arabella Stuart 1 v. — Rose d'Albret 1 v. — Arrah Neil 1 v. — Agincourt 1 v. — The Smuggler 1 v. — The Step-Mother 2 v. — Beauchamp 1 v. — Heidelberg 1 v. — The Gipsy 1 v. — The Castle of Ehrenstein 1 v. — Darnley 1 v. — Russell 2 v. — The Convict 2 v. — Sir Theodore Broughton 2 v.

James, Henry (Am.).

The American 2 v. — The Europeans 1 v. — Daisy Miller; An International Episode; Four Meetings 1 v. — Roderick Hudson 2 v. — The Madonna of the Future, etc. 1 v. — Eugene Pickering, etc. 1 v. — Confidence 1 v. — Washington Square, etc. 2 v. — The Portrait of a Lady 3 v. — Foreign Parts 1 v. — French Poets and Novelists 1 v. — The Siege of London; The Point of View; A Passionate Pilgrim 1 v. — Portraits of Places 1 v. — A Little Tour in France 1 v. — The Finer Grain 1 v. — The Outcry 1 v.

James, Winifred.

Bachelor Betty 1 v.

Jeaffreson, J. Cordy.

A Book about Doctors 2 v. — A Woman in spite of Herself 2 v. — The Real Lord Byron 3 v.

Jenkin, Mrs. Charles, † 1885.

"Who Breaks—Pays" 1 v. — Skirmishing 1 v. — Once and Again 2 v. — Two French Marriages 2 v. — Within an Ace 1 v. — Jupiter's Daughters 1 v.

Jenkins, Edward.

Ginx's Baby, his Birth and other Misfortunes; Lord Bantam 2 v.

"Jennie of 'The Prince's,'" Author of: *vide* B. H. Buxton.

Jerome, K. Jerome.

The Idle Thoughts of an Idle Fellow 1 v. — Diary of a Pilgrimage, and Six Essays 1 v. — Novel Notes 1 v. — Sketches in Lavender, Blue and Green 1 v. — The Second Thoughts of an Idle Fellow 1 v. — Three Men on the Bummel 1 v. — Paul Kelver 2 v. — Tea-Table Talk 1 v. — Tommy and Co. 1 v. — Idle Ideas in 1905 1 v. — The Passing of the Third Floor Back 1 v. — The Angel and the Author—and Others 1 v. — They and I, 1 v.

Jerrold, Douglas, † 1857.

History of St. Giles and St. James 2 v. — Men of Character 2 v.

"John Halifax, Gentleman," Author of: *vide* Mrs. Craik.

Johnny Ludlow: *vide* Mrs. Henry Wood.

Johnson, Samuel, † 1784.

Lives of the English Poets 2 v.

Jolly, Emily.
Colonel Dacre 2 v.

"Joshua Davidson," Author of: *vide* Mrs. E. Lynn Linton.

Kavanagh, Miss Julia, † 1877.
Nathalie 2 v. — Daisy Burns 2 v. — Grace Lee 2 v. — Rachel Gray 1 v. — Adèle 3 v. — A Summer and Winter in the Two Sicilies 2 v. — Seven Years, and other Tales 2 v. — French Women of Letters 1 v. — English Women of Letters 1 v. — Queen Mab 2 v. — Beatrice 2 v. — Sybil's Second Love 2 v. — Dora 2 v. — Silvia 2 v. — Bessie 2 v. — John Dorrien 3 v. — Two Lilies 2 v. — Forget-me-nots 2 v. — *Vide* Series for the Young, p. 29.

Keary, Annie, † 1879.
Oldbury 2 v. — Castle Daly 2 v.

Keary, C. F.
The Mount 1 v.

Keeling, D'Esterre-: *v.* Esterre.

Kempis, Thomas a.
The Imitation of Christ. Translated from the Latin by W. Benham, B.D. 1 v.

Kimball, Richard B. (Am.), † 1892.
Saint Leger 1 v. — Romance of Student Life Abroad 1 v. — Undercurrents 1 v. — Was he Successful? 1 v. — To-Day in New York 1 v.

Kinglake, A. W., † 1891.
Eothen 1 v. — The Invasion of the Crimea 14 v.

Kingsley, Charles, † 1875.
Yeast 1 v. — Westward ho! 2 v. — Two Years ago 2 v. — Hypatia 2 v. — Alton Locke 1 v. — Hereward the Wake 2 v. — At Last 2 v. — His Letters and Memories of his Life, edited by his Wife 2 v.

Kingsley, Henry, † 1876.
Ravenshoe 2 v. — Austin Elliot 1 v. — Geoffry Hamlyn 2 v. — The Hillyars and the Burtons 2 v. — Leighton Court 1 v. — Valentin 1 v. — Oakshott Castle 1 v. — Reginald Hetherege 2 v. — The Grange Garden 2 v.

Kinross, Albert.
An Opera and Lady Grasmere 1 v.

Kipling, Rudyard.
Plain Tales from the Hills 1 v. — The Second Jungle Book 1 v. — The Seven Seas 1 v. — "Captains Courageous" 1 v. — The Day's Work 1 v. — A Fleet in Being 1 v. — Stalky & Co. 1 v. — From Sea to Sea 2 v. — The City of Dreadful Night 1 v. — Kim 1 v. — Just So Stories 1 v. — The Five Nations 1 v. — Traffics and Discoveries 1 v. — Puck of Pook's Hill 1 v. — Actions and Reactions 1 v. — Rewards and Fairies 1 v.

Laffan, May.
Flitters, Tatters, and the Counsellor 1 v.

Lamb, Charles, † 1834.
The Essays of Elia and Eliana 1 v.

Lang, Andrew: *vide* H. Rider Haggard.

Langdon, Mary (Am.).
Ida May 1 v.

"Last of the Cavaliers, the," Author of (Miss Piddington).
The Last of the Cavaliers 2 v. — The Gain of a Loss 2 v.

Łaszowska, Mme de: *vide* E. Gerard.

Laurence, George Alfred: *vide* "Guy Livingstone."

Lawless, the Hon. Emily.
Hurrish 1 v.

Lee, Holme: *vide* Harriet Parr.

Lee, Vernon.
Pope Jacynth, etc. 1 v. — Genius Loci, and The Enchanted Woods 1 v. — Hortus Vitae, and Limbo 1 v. — The Spirit of Rome, and Laurus Nobilis 1 v. — Vanitas 1 v.

Le Fanu, J. S., † 1873.
Uncle Silas 2 v. — Guy Deverell 2 v.

Lemon, Mark, † 1870.
Wait for the End 2 v. — Loved at Last 2 v. — Falkner Lyle 2 v. — Leyton Hall, and other Tales 2 v. — Golden Fetters 2 v.

Lever, Charles, † 1872.
The O'Donoghue 1 v. — The Knight of Gwynne 3 v. — Arthur O'Leary 2 v. — Harry Lorrequer 2 v. — Charles O'Malley 3 v. — Tom Burke of "Ours" 3 v. — Jack Hinton 2 v. — The Daltons 4 v. — The Dodd Family Abroad 3 v. — The Martins of Cro' Martin 3 v. — The Fortunes of Glencore 2 v. — Roland Cashel 3 v. — Davenport Dunn 3 v. — Confessions of Con Cregan 2 v. — One of Them 2 v. — Maurice Tiernay 2 v. — Sir Jasper Carew 2 v. — Barrington 2 v. — A Day's Ride 2 v. — Luttrell of Arran 2 v. — Tony Butler 2 v. — Sir Brook Fossbrooke 2 v. — The Bramleighs of Bishop's Folly 2 v. — A Rent in a Cloud 1 v. — That Boy of Norcott's 1 v. — St. Patrick's Eve; Paul

Gosslett's Confessions 1 v. — Lord Kilgobbin 2 v.

Levett-Yeats, S.

The Honour of Savelli 1 v. — The Chevalier d'Auriac 1 v. — The Traitor's Way 1 v. — The Lord Protector 1 v. — Orrain 1 v.

Lewes, G. H., † 1878.

Ranthorpe 1 v. — The Physiology of Common Life 2 v. — On Actors and the Art of Acting 1 v.

Linton, Mrs. E. Lynn, † 1898.

The true History of Joshua Davidson 1 v. — Patricia Kemball 2 v. — The Atonement of Leam Dundas 2 v. — The World well Lost 2 v. — Under which Lord? 2 v. — With a Silken Thread, and other Stories 1 v. — Todhunters' at Loanin' Head, and other Stories 1 v. — "My Love!" 2 v. — The Girl of the Period, and other Social Essays 1 v. — Ione 2 v.

Lockhart, Laurence W. M., † 1882.

Mine is Thine 2 v.

Loftus, Lord Augustus.

Diplomatic Reminiscences 1837 - 1862 (with Portrait) 2 v.

London, Jack (Am.).

Burning Daylight 1 v.

Longard, Mme de: *v.* D. Gerard.

Longfellow, Henry Wadsworth (Am.), † 1882.

Poetical Works (with Portrait) 3 v. — The Divine Comedy of Dante Alighieri 3 v. — The New-England Tragedies 1 v. — The Divine Tragedy 1 v. — Flower-de-Luce, and Three Books of Song 1 v. — The Masque of Pandora, and other Poems 1 v.

Lonsdale, Margaret.

Sister Dora (with Portrait) 1 v.

Lorimer, George Horace (Am.).

Letters from a Self-Made Merchant to his Son 1 v. — Old Gorgon Graham 1 v. — Jack Spurlock, Prodigal 1 v.

"Lost Battle, a," Author of.

A Lost Battle 2 v.

Lowndes, Mrs. Belloc.

The Uttermost Farthing 1 v. — Studies in Wives 1 v. — When No Man Pursueth 1 v. — Jane Oglander 1 v.

Lubbock, Sir John (Lord Avebury).

The Pleasures of Life 1 v. — The Beauties of Nature (with Illustrations) 1 v. — The Use of Life 1 v. — Scenery of Switzerland (with Illustrations) 2 v. — Essays and Addresses 1900-1903 1 v. — On Peace and Happiness 1 v.

"Lutfullah": *vide* Eastwick.

Lyall, Edna, † 1903.

We Two 2 v. — Donovan 2 v. — In the Golden Days 2 v. — Knight-Errant 2 v. — Won by Waiting 2 v. — Wayfaring Men 2 v. — Hope the Hermit 2 v. — Doreen 2 v. — In Spite of All 2 v. — The Hinderers 1 v.

Lytton, Lord: *vide* E. Bulwer.

Lytton, Robert Lord (Owen Meredith), † 1891.

Poems 2 v. — Fables in Song 2 v.

Maartens, Maarten.

The Sin of Joost Avelingh 1 v. — An Old Maid's Love 2 v. — God's Fool 2 v. — The Greater Glory 2 v. — My Lady Nobody 2 v. — Her Memory 1 v. — Some Women I have known 1 v. — My Poor Relations 2 v. — Dorothea 2 v. — The Healers 2 v. — The Woman's Victory, and Other Stories 2 v. — The New Religion 2 v. — Brothers All 1 v. — The Price of Lis Doris 2 v. — Harmen Pols: Peasant 1 v.

McAulay, Allan (Am.): *vide* Kate Douglas Wiggin.

Macaulay, Lord, † 1859.

History of England (with Portrait) 10 v. — Critical and Historical Essays 5 v. — Lays of Ancient Rome 1 v. — Speeches 2 v. — Biographical Essays 1 v. — William Pitt, Atterbury 1 v. — (See also Trevelyan).

McCarthy, Justin.

The Waterdale Neighbours 2 v. — Dear Lady Disdain 2 v. — Miss Misanthrope 2 v. — A History of our Own Times 5 v. — Donna Quixote 2 v. — A Short History of our Own Times 2 v. — A History of the Four Georges. Vols. 1 & 2. — A History of our Own Times. Vols. 6 & 7 (supplemental). — A History of the Four Georges and of William IV. Vols. 3, 4 & 5 (supplemental). — A Short History of our Own Times. Vol. 3 (supplemental).

Mac Donald, George, † 1905.

Alec Forbes of Howglen 2 v. — Annals of a Quiet Neighbourhood 2 v. — David Elginbrod 2 v. — The Vicar's Daughter 2 v. — Malcolm 2 v. — St. George and St. Michael 2 v. — The Marquis of Lossie 2 v. — Sir Gibbie 2 v. — Mary

Marston 2 v. — The Gifts of the Child Christ, and other Tales 1 v. — The Princess and Curdie 1 v.

Mackarness, Mrs., † 1881.
Sunbeam Stories 1 v. — A Peerless Wife 2 v. — A Mingled Yarn 2 v.

Mackay, Eric, † 1898.
Love Letters of a Violinist, and other Poems 1 v.

McKnight, Charles (Am.), † 1881.
Old Fort Duquesne 2 v.

Maclaren, Ian, † 1907.
Beside the Bonnie Brier Bush 1 v. — The Days of Auld Langsyne 1 v. — His Majesty Baby 1 v.

Macleod, Fiona, † 1905.
Wind and Wave 1 v. — The Sunset of Old Tales 1 v.

Macleod, Norman, † 1872.
The Old Lieutenant and his Son 1 v.

Macpherson, James, † 1796: *vide* Ossian.

Macquoid, Mrs.
Patty 2 v. — Miriam's Marriage 2 v. — Pictures across the Channel 2 v. — Too Soon 1 v. — My Story 2 v. — Diane 2 v. — Beside the River 2 v. — A Faithful Lover 2 v.

"Mademoiselle Mori," Author of (Miss Roberts).
Mademoiselle Mori 2 v. — Denise 1 v. — Madame Fontenoy 1 v. — On the Edge of the Storm 1 v. — The Atelier du Lys 2 v. — In the Olden Time 2 v.

Mahon, Lord: *vide* Stanhope.

Maine, E. S.
Scarscliff Rocks 2 v.

Malet, Sir Edward.
Shifting Scenes 1 v.

Malet, Lucas (Mrs. Mary St. Leger Harrison).
Colonel Enderby's Wife 2 v. — The History of Sir Richard Calmady 3 v. — The Far Horizon 2 v. — The Score 1 v. — Adrian Savage 2 v.

Malmesbury, the Earl of.
Memoirs of an Ex-Minister 3 v.

Mann, Mary E.
A Winter's Tale 1 v. — The Cedar Star 1 v.

Mansfield, Robert Blachford.
The Log of the Water Lily 1 v.

Mark Twain: *vide* Twain.

"Marmorne," Author of: *vide* P. G. Hamerton.

Marryat, Capt., † 1848.
Jacob Faithful (with Portrait) 1 v. — Percival Keene 1 v. — Peter Simple 1 v. — Japhet in Search of a Father 1 v. — Monsieur Violet 1 v. — The Settlers in Canada 1 v. — The Mission 1 v. — The Privateer's-Man 1 v. — The Children of the New-Forest 1 v. — Valerie 1 v. — Mr. Midshipman Easy 1 v. — The King's Own 1 v.

Marryat, Florence, † 1899.
Love's Conflict 2 v. — For Ever and Ever 2 v. — The Confessions of Gerald Estcourt 2 v. — Nelly Brooke 2 v. — Véronique 2 v. — Petronel 2 v. — Her Lord and Master 2 v. — The Prey of the Gods 1 v. — Life and Letters of Captain Marryat 1 v. — Mad Dumaresq 2 v. — No Intentions 2 v. — Fighting the Air 2 v. — A Star and a Heart; An Utter Impossibility 1 v. — The Poison of Asps, and other Stories 1 v. — A Lucky Disappointment, and other Stories 1 v. — "My own Child" 2 v. — Her Father's Name 2 v. — A Harvest of Wild Oats 2 v. — A Little Stepson 1 v. — Written in Fire 2 v. — Her World against a Lie 2 v. — A Broken Blossom 2 v. — The Root of all Evil 2 v. — The Fair-haired Alda 2 v. — With Cupid's Eyes 2 v. — My Sister the Actress 2 v. — Phyllida 2 v. — How they loved Him 2 v. — Facing the Footlights (with Portrait) 2 v. — A Moment of Madness, and other Stories 1 v. — The Ghost of Charlotte Cray, and other Stories 1 v. — Peeress and Player 2 v. — Under the Lilies and Roses 2 v. — The Heart of Jane Warner 2 v. — The Heir Presumptive 2 v. — The Master Passion 2 v. — Spiders of Society 2 v. — Driven to Bay 2 v. — A Daughter of the Tropics 2 v. — Gentleman and Courtier 2 v. — On Circumstantial Evidence 2 v. — Mount Eden. A Romance 2 v. — Blindfold 2 v. — A Scarlet Sin 1 v. — A Bankrupt Heart 2 v. — The Spirit World 1 v. — The Beautiful Soul 1 v. — At Heart a Rake 2 v. — The Strange Transfiguration of Hannah Stubbs 1 v. — The Dream that Stayed 2 v. — A Passing Madness 1 v. — The Blood of the Vampire 1 v. — A Soul on Fire 1 v. — Iris the Avenger 1 v.

Marsh, Mrs. Anne, † 1874.
Ravenscliffe 2 v. — Emilia Wyndham 2 v. — Castle Avon 2 v. — Aubrey 2 v. —

The Heiress of Haughton 2 v. — Evelyn Marston 2 v. — The Rose of Ashurst 2 v.

Marshall, Mrs. Emma, † 1899.

Mrs. Mainwaring's Journal 1 v. — Benvenuta 1 v. — Lady Alice 1 v. — Dayspring 1 v. — Life's Aftermath 1 v. — In the East Country 1 v. — No. XIII; or, The Story of the Lost Vestal 1 v. — In Four Reigns 1 v. — On the Banks of the Ouse 1 v. — In the City of Flowers 1 v. — Alma 1 v. — Under Salisbury Spire 1 v. — The End Crowns All 1 v. — Winchester Meads 1 v. — Eventide Light 1 v. — Winifrede's Journal 1 v. — Bristol Bells 1 v. — In the Service of Rachel Lady Russell 1 v. — A Lily among Thorns 1 v. — Penshurst Castle 1 v. — Kensington Palace 1 v. — The White King's Daughter 1 v. — The Master of the Musicians 1 v. — An Escape from the Tower 1 v. — A Haunt of Ancient Peace 1 v. — Castle Meadow 1 v. — In the Choir of Westminster Abbey 1 v. — The Young Queen of Hearts 1 v. — Under the Dome of St. Paul's 1 v. — The Parson's Daughter 1 v.

Mason, A. E. W.

The Four Feathers 2 v. — Miranda of the Balcony 1 v. — The Courtship of Morrice Buckler 2 v. — The Truants 2 v. — The Watchers 1 v. — Running Water 1 v. — The Broken Road 1 v. — At the Villa Rose 1 v.

Mathers, Helen (Mrs. Henry Reeves).

"Cherry Ripe!" 2 v. — "Land o' the Leal" 1 v. — My Lady Green Sleeves 2 v. — As he comes up the Stair, etc. 1 v. — Sam's Sweetheart 2 v. — Eyre's Acquittal 2 v. — Found Out 1 v. — Murder or Manslaughter? 1 v. — The Fashion of this World (80 Pf.) — Blind Justice, and "Who, being dead, yet Speaketh" 1 v. — What the Glass Told, and A Study of a Woman 1 v. — Bam Wildfire 2 v. — Becky 2 v. — Cinders 1 v. — "Honey" 1 v. — Griff of Griffithscourt 1 v. — The New Lady Teazle, and Other Stories and Essays 1 v. — The Ferryman 1 v. — Tally Ho! 2 v. — Pigskin and Petticoat 2 v. — Gay Lawless 1 v. — Love the Thief 1 v.

Maurice, Colonel.

The Balance of Military Power in Europe 1 v.

Maurier, George du, † 1896.

Trilby 2 v. — The Martian 2 v.

Maxwell, Mrs.: *v.* Miss Braddon.

Maxwell, W. B.

The Ragged Messenger 2 v. — The Guarded Flame 2 v. — Mrs. Thompson 1 v. — The Rest Cure 1 v.

"Mehalah": *v.* Baring-Gould.

Melville, George J. Whyte, † 1878.

Kate Coventry 1 v. — Holmby House 2 v. — Digby Grand 1 v. — Good for Nothing 2 v. — The Queen's Maries 2 v. — The Gladiators 2 v. — The Brookes of Bridlemere 2 v. — Cerise 2 v. — The Interpreter 2 v. — The White Rose 2 v. — M. or N. 1 v. — Contraband 1 v. — Sarchedon 2 v — Uncle John 2 v. — Katerfelto 1 v. — Sister Louise 1 v. — Rosine 1 v. — Roys' Wife 2 v. — Black but Comely 2 v. — Riding Recollections 1 v.

Memorial Volumes: *vide* Five Centuries (vol. 500); The New Testament (vol. 1000); Henry Morley (vol. 2000).

Meredith, George, † 1909.

The Ordeal of Richard Feverel 2 v. — Beauchamp's Career 2 v. — The Tragic Comedians 1 v. — Lord Ormont and his Aminta 2 v. — The Amazing Marriage 2 v. — The Egoist 2 v.

Meredith, Owen: *vide* Robert Lord Lytton.

Merrick, Leonard.

The Man who was good 1 v. — This Stage of Fools 1 v. — Cynthia 1 v. — One Man's View 1 v. — The Actor-Manager 1 v. — The Worldlings 1 v. — When Love flies out o' the Window 1 v. — Conrad in Quest of His Youth 1 v. — The Quaint Companions 1 v. — Whispers about Women 1 v. — The House of Lynch 1 v. — The Man who Understood Women, etc. 1 v. — All the World Wondered, etc. 1 v. — The Position of Peggy Harper 1 v.

Merriman, Henry Seton, † 1903.

Young Mistley 1 v. — Prisoners and Captives 2 v. — From One Generation to Another 1 v. — With Edged Tools 2 v. — The Sowers 2 v. — Flotsam 1 v. — In Kedar's Tents 1 v. — Roden's Corner 1 v. — The Isle of Unrest 1 v. — The Velvet Glove 1 v. — The Vultures 1 v. — Barlasch of the Guard 1 v. — Tomaso's Fortune, and Other Stories 1 v. — The Last Hope 2 v.

Merriman, H. S., & S. G. Tallentyre.

The Money-Spinner, etc. 1 v.

Milne, James.

The Epistles of Atkins 1 v.

Milton, John, † 1674.

Poetical Works 1 v.

"Miss Molly," Author of.

Geraldine Hawthorne 1 v.

"Molly Bawn," Author of: *vide* Mrs. Hungerford.

Montgomery, Florence.

Misunderstood 1 v. — Thrown Together 2 v. — Thwarted 1 v. — Wild Mike 1 v. — Seaforth 2 v. — The Blue Veil 1 v. — Transformed 1 v. — The Fisherman's Daughter, etc. 1 v. — Colonel Norton 2 v. — Prejudged 1 v. — An Unshared Secret, and Other Tales 1 v.

Moore, Frank Frankfort.

"I Forbid the Banns" 2 v. — A Gray Eye or So 2 v. — One Fair Daughter 2 v. — They Call it Love 2 v. — The Jessamy Bride 1 v. — The Millionaires 1 v. — Nell Gwyn—Comedian 1 v.—A Damsel or Two 1 v. — Castle Omeragh 2 v. — Shipmates in Sunshine 2 v. — The Original Woman 1 v. — The White Causeway 1 v.— The Artful Miss Dill 1 v. — The Marriage Lease 1 v. — An Amateur Adventuress 1 v. — Priscilla and Charybdis 1 v. — The Food of Love 1 v.—The Laird of Craig Athol 1 v. — The Marriage of Barbara 1 v.

Moore, George.

Celibates 1 v. — Evelyn Innes 2 v. — Sister Teresa 2 v.—The Untilled Field 1 v. — Confessions of a Young Man 1 v. — The Lake 1 v. — Memoirs of my Dead Life 1 v.

Moore, Thomas, † 1852.

Poetical Works (with Portrait) 5 v.

Morgan, Lady, † 1859.

Memoirs 3 v.

Morley, Henry, † 1894.

Of English Literature in the Reign of Victoria. With Facsimiles of the Signatures of Authors in the Tauchnitz Edition (v. 2000, published 1881) 1 v.

Morris, William.

A Selection from his Poems. Edited with a Memoir by F. Hueffer 1 v.

Morrison, Arthur.

Tales of Mean Streets 1 v. — A Child of the Jago 1 v. — To London Town 1 v. — Cunning Murrell 1 v. — The Hole in the Wall 1 v. — The Green Eye of Goona 1 v. — Divers Vanities 1 v. — Green Ginger 1 v.

Muirhead, James Fullarton.

The Land of Contrasts 1 v.

Mulock, Miss: *vide* Mrs. Craik.

Murray, David Christie.

Rainbow Gold 2 v.

Murray, Grenville: *v.* Grenville.

"My Little Lady," Author of: *vide* E. Frances Poynter.

New Testament, the.

The Authorised English Version, with Introduction and Various Readings from the three most celebrated Manuscripts of the Original Text, by Constantine Tischendorf (vol. 1000, published 1869) 1 v.

Newby, Mrs. C. J.

Common Sense 2 v.

Newman, Dr. J. H. (Cardinal Newman), † 1890.

Callista 1 v.

Nicholls, Mrs.: *vide* Currer Bell.

"Nina Balatka," Author of: *vide* Anthony Trollope.

"No Church," Author of (F. Robinson).

No Church 2 v. — Owen:—a Waif 2 v.

Noel, Lady Augusta.

From Generation to Generation 1 v. — Hithersea Mere 2 v.

Norris, Frank (Am.), † 1902.

The Octopus 2 v. — The Pit 2 v.

Norris, W. E.

My Friend Jim 1 v. — A Bachelor's Blunder 2 v. — Major and Minor 2 v. — The Rogue 2 v. — Miss Shafto 2 v. — Mrs. Fenton 1 v. — Misadventure 2 v. — Saint Ann's 1 v. — A Victim of Good Luck 1 v. — The Dancer in Yellow 1 v. — Clarissa Furiosa 2 v. — Marietta's Marriage 2 v. — The Fight for the Crown 1 v.—The Widower 1 v.—Giles Ingilby 1 v. — The Flower of the Flock 1 v. — His Own Father 1 v.—The Credit of the County 1 v. — Lord Leonard the Luckless 1 v. — Nature's Comedian 1 v.—Nigel's Vocation 1 v. — Barham of Beltana 1 v. — Harry and Ursula 1 v. — The Square Peg 1 v. — Pauline 1 v. — The Perjurer 1 v. — Not Guilty 1 v. — Vittoria Victrix 1 v.

Norton, Hon. Mrs., † 1877.

Stuart of Dunleath 2 v. — Lost and Saved 2 v. — Old Sir Douglas 2 v.

"Not Easily Jealous," Author of (Miss Iza Hardy).

Not Easily Jealous 2 v.

"Novels and Tales": *vide* "Household Words."

O'Conor Eccles, Charlotte (Hal Godfrey).

The Rejuvenation of Miss Semaphore 1 v. — The Matrimonial Lottery 1 v.

Oldmeadow, Ernest.

Susan 1 v.

Oliphant, Laurence, † 1888.

Altiora Peto 2 v. — Masollam 2 v.

Oliphant, Mrs., † 1897.

The Last of the Mortimers 2 v. — Mrs. Margaret Maitland 1 v. — Agnes 2 v. — Madonna Mary 2 v. — The Minister's Wife 2 v. — The Rector and the Doctor's Family 1 v. — Salem Chapel 2 v. — The Perpetual Curate 2 v. — Miss Marjoribanks 2 v. — Ombra 2 v. — Memoir of Count de Montalembert 2 v. — May 2 v. — Innocent 2 v. — For Love and Life 2 v. — A Rose in June 1 v. — The Story of Valentine and his Brother 2 v. — Whiteladies 2 v. — The Curate in Charge 1 v. — Phœbe, Junior 2 v. — Mrs. Arthur 2 v. — Carità 2 v. — Young Musgrave 2 v. — The Primrose Path 2 v. — Within the Precincts 3 v. — The Greatest Heiress in England 2 v. — He that will not when he may 2 v. — Harry Joscelyn 2 v. — In Trust 2 v. — It was a Lover and his Lass 3 v. — The Ladies Lindores 3 v. — Hester 3 v. — The Wizard's Son 3 v. — A Country Gentleman and his Family 2 v. — Neighbours on the Green 1 v. — The Duke's Daughter 1 v. — The Fugitives 1 v. — Kirsteen 2 v. — Life of Laurence Oliphant and of Alice Oliphant, his Wife 2 v. — The Little Pilgrim in the Unseen 1 v. — The Heir Presumptive and the Heir Apparent 2 v. — The Sorceress 2 v. — Sir Robert's Fortune 2 v. — The Ways of Life 1 v. — Old Mr. Tredgold 2 v.

"One who has kept a Diary": *vide* George W. E. Russell.

Orczy, Baroness.

Petticoat Government 1 v. — The Scarlet Pimpernel 1 v. — I will Repay 1 v. — The Elusive Pimpernel 1 v.

Osbourne, Lloyd (Am.).

Baby Bullet 1 v. — Wild Justice 1 v. — The Motormaniacs 1 v. — Harm's Way 1 v.

Ossian.

The Poems of Ossian. Translated by James Macpherson 1 v.

Ouida, † 1908.

Idalia 2 v. — Tricotrin 2 v. — Puck 2 v. — Chandos 2 v. — Strathmore 2 v. — Under two Flags 2 v. — Folle-Farine 2 v. — A Leaf in the Storm; A Dog of Flanders; A Branch of Lilac; A Provence Rose 1 v. — Cecil Castlemaine's Gage, and other Novelettes 1 v. — Madame la Marquise, and other Novelettes 1 v. — Pascarèl 2 v. — Held in Bondage 2 v. — Two little Wooden Shoes 1 v. — Signa (with Portrait) 3 v. — In a Winter City 1 v. — Ariadnê 2 v. — Friendship 2 v. — Moths 3 v. — Pipistrello, and other Stories 1 v. — A Village Commune 2 v. — In Maremma 3 v. — Bimbi 1 v. — Wanda 3 v. — Frescoes and other Stories 1 v. — Princess Napraxine 3 v. — Othmar 3 v. — A Rainy June (60 Pf.). Don Gesualdo (60 Pf.). — A House Party 1 v. — Guilderoy 2 v. — Syrlin 3 v. — Ruffino, and other Stories 1 v. — Santa Barbara, etc. 1 v. — Two Offenders 1 v. — The Silver Christ, etc. 1 v. — Toxin, and other Papers 1 v. — Le Selve, and Tonia 1 v. — The Massarenes 2 v. — An Altruist, and Four Essays 1 v. — La Strega, and other Stories 1 v. — The Waters of Edera 1 v. — Street Dust, and Other Stories 1 v. — Critical Studies 1 v. — Helianthus 2 v.

"Outcasts, the," Author of: *vide* "Roy Tellet."

Pain, Barry.

The Exiles of Faloo 1 v.

Parker, Sir Gilbert.

The Battle of the Strong 2 v. — Donovan Pasha, & Some People of Egypt 1 v. — The Seats of the Mighty 2 v. — The Weavers 2 v.

Parr, Harriet (Holme Lee), † 1900.

Basil Godfrey's Caprice 2 v. — For Richer, for Poorer 2 v. — The Beautiful Miss Barrington 2 v. — Her Title of Honour 1 v. — Echoes of a Famous Year 1 v. — Katherine's Trial 1 v. — The Vicissitudes of Bessie Fairfax 2 v. — Ben Milner's Wooing 1 v. — Straightforward 2 v. — Mrs. Denys of Cote 2 v. — A Poor Squire 1 v.

Parr, Mrs.

Dorothy Fox 1 v. — The Prescotts of Pamphillon 2 v. — The Gosau Smithy, etc. 1 v. — Robin 2 v. — Loyalty George 2 v.

Paston, George.

A Study in Prejudices 1 v. — A Fair Deceiver 1 v.

Pasture, Mrs. Henry de la.

The Lonely Lady of Grosvenor Square 1 v. — The Grey Knight 1 v. — Catherine's Child 1 v. — Master Christopher 2 v.

Paul, Mrs.: *vide* Author of "Still Waters."

"Paul Ferroll," Author of (Mrs. Caroline Clive), † 1873.

Paul Ferroll 1 v. — Year after Year 1 v. — Why Paul Ferroll killed his Wife 1 v.

Payn, James, † 1898.

Found Dead 1 v. — Gwendoline's Harvest 1 v. — Like Father, like Son 2 v. — Not Wooed, but Won 2 v. — Cecil's Tryst 1 v. — A Woman's Vengeance 2 v. — Murphy's Master 1 v. — In the Heart of a Hill, and other Stories 1 v. — At Her Mercy 2 v. — The Best of Husbands 2 v. — Walter's Word 2 v. — Halves 2 v. — Fallen Fortunes 2 v. — What He cost Her 2 v. — By Proxy 2 v. — Less Black than we're Painted 2 v. — Under one Roof 2 v. — High Spirits 1 v. — High Spirits (*Second Series*) 1 v. — A Confidential Agent 2 v. — From Exile 2 v. — A Grape from a Thorn 2 v. — Some Private Views 1 v. — For Cash Only 2 v. — Kit: A Memory 2 v. — The Canon's Ward (with Portrait) 2 v. — Some Literary Recollections 1 v. — The Talk of the Town 1 v. — The Luck of the Darrells 2 v. — The Heir of the Ages 2 v. — Holiday Tasks 1 v. — Glow-Worm Tales (*First Series*) 1 v. — Glow-Worm Tales (*Second Series*) 1 v. — A Prince of the Blood 2 v. — The Mystery of Mirbridge 2 v. — The Burnt Million 2 v. — The Word and the Will 2 v. — Sunny Stories, and some Shady Ones 1 v. — A Modern Dick Whittington 2 v. — A Stumble on the Threshold 2 v. — A Trying Patient 1 v. — Gleams of Memory, and The Eavesdropper 1 v. — In Market Overt 1 v. — The Disappearance of George Driffell, and other Tales 1 v. — Another's Burden etc. 1 v. — The Backwater of Life, or Essays of a Literary Veteran 1 v.

Peard, Frances Mary.

One Year 2 v. — The Rose-Garden 1 v. — Unawares 1 v. — Thorpe Regis 1 v. — A Winter Story 1 v. — A Madrigal, and other Stories 1 v. — Cartouche 1 v. — Mother Molly 1 v. — Schloss and Town 2 v. — Contradictions 2 v. — Near Neighbours 1 v. — Alicia Tennant 1 v. — Madame's Granddaughter 1 v. — Donna Teresa 1 v. — Number One and Number Two 1 v. — The Ring from Jaipur 1 v. — The Flying Months 1 v.

Pemberton, Max.

The Impregnable City 1 v. — A Woman of Kronstadt 1 v. — The Phantom Army 1 v. — The Garden of Swords 1 v. — The Footsteps of a Throne 1 v. — Pro Patriâ 1 v. — The Giant's Gate 2 v. — I crown thee King 1 v. — The House under the Sea 1 v. — The Gold Wolf 1 v. — Doctor Xavier 1 v. — Red Morn 1 v. — Beatrice of Venice 2 v. — Mid the Thick Arrows 2 v. — My Sword for Lafayette 1 v. — The Lady Evelyn 1 v. — The Diamond Ship 1 v. — The Lodestar 1 v. — Wheels of Anarchy 1 v. — Love the Harvester 1 v. — The Adventures of Captain Jack 1 v. — White Walls 1 v. — The Show Girl 1 v.

Percy, Bishop Thomas, † 1811.

Reliques of Ancient English Poetry 3 v.

Perrin, Alice.

Idolatry 1 v. — The Charm 1 v.

Philips, F. C.

As in a Looking Glass 1 v. — The Dean and his Daughter 1 v. — Lucy Smith 1 v. — A Lucky Young Woman 1 v. — Jack and Three Jills 1 v. — Little Mrs. Murray 1 v. — Young Mr. Ainslie's Courtship 1 v. — Social Vicissitudes 1 v. — Extenuating Circumstances, and A French Marriage 1 v. — More Social Vicissitudes 1 v. — Constance 2 v. — That Wicked Mad'moiselle, etc. 1 v. — A Doctor in Difficulties, etc. 1 v. — Black and White 1 v. — "One Never Knows" 2 v. — Of Course 1 v. — Miss Ormerod's Protégé 1 v. — My little Husband 1 v. — Mrs. Bouverie 1 v. — A Question of Colour, and other Stories 1 v. — A Devil in Nun's Veiling 1 v. — A Full Confession, and other Stories 1 v. — The Luckiest of Three 1 v. — Poor Little Bella 1 v. — Eliza Clarke, Governess, and Other Stories 1 v. — Marriage, etc. 1 v. — School-girls of To-day, etc. 1 v. — If Only, etc. 1 v. — An Unfortunate Blend 1 v. — A Barrister's Courtship 1 v.

Philips, F. C. & Percy Fendall.

A Daughter's Sacrifice 1 v. — Margaret Byng 1 v. — Disciples of Plato 1 v. — A Honeymoon—and After 1 v.

Philips, F. C. & C. J. Wills.

The Fatal Phryne 1 v. — The Scudamores 1 v. — A Maiden Fair to See 1 v. — Sybil Ross's Marriage 1 v.

Philips, F. C. & A. R. T.

Life 1 v.

Phillpotts, Eden.

Lying Prophets 2 v. — The Human Boy 1 v. — Sons of the Morning 2 v. — The Good Red Earth 1 v. — The Striking Hours 1 v. — The Farm of the Dagger 1 v. —

The Golden Fetich 1 v. — The Whirlwind 2 v. — The Human Boy Again 1 v.

Phillpotts, E. & Arnold Bennett.
The Sinews of War 1 v. — The Statue 1 v.

Piddington, Miss: *vide* Author of "The Last of the Cavaliers."

Poe, Edgar Allan (Am.), † 1849.
Poems and Essays, edited with a new Memoir by John H. Ingram 1 v. — Tales, edited by John H. Ingram 1 v.

Pope, Alexander, † 1744.
Select Poetical Works (with Portrait) 1 v.

Poynter, Miss E. Frances.
My Little Lady 2 v. — Ersilia 2 v. — Among the Hills 1 v. — Madame de Presnel 1 v.

Praed, Mrs. Campbell.
Zéro 1 v. — Affinities 1 v. — The Head Station 2 v.

Prentiss, Mrs. E. (Am.), † 1878.
Stepping Heavenward 1 v.

Prince Consort, the, † 1861.
His Principal Speeches and Addresses (with Portrait) 1 v.

Pryce, Richard.
Miss Maxwell's Affections 1 v. — The Quiet Mrs. Fleming 1 v. — Time and the Woman 1 v.

Pym, H. N.: *vide* Caroline Fox.

Quiller-Couch, A. T. ("Q").
Noughts and Crosses 1 v. — I Saw Three Ships 1 v. — Dead Man's Rock 1 v. — Ia and other Tales 1 v. — The Ship of Stars 1 v. — The Adventures of Harry Revel 1 v. — Fort Amity 1 v. — Shakespeare's Christmas, and Other Stories 1 v. — The Mayor of Troy 1 v. — Merry-Garden, and Other Stories 1 v. — Brother Copas 1 v.

Quincey: *vide* De Quincey.

Rae, W. Fraser, † 1905.
Westward by Rail 1 v. — Miss Bayle's Romance 2 v. — The Business of Travel 1 v

Raimond, C. E. (Miss Robins) (Am.).
The Open Question 2 v. — The Magnetic North 2 v. — A Dark Lantern 2 v. — The Convert 2 v. — The Florentine Frame 1 v.

"Rajah's Heir, the," Author of.
The Rajah's Heir 2 v.

Reade, Charles, † 1884.
"It is never too late to mend" 2 v. — "Love me little, love me long" 1 v. — The Cloister and the Hearth 2 v. — Hard Cash 3 v. — Put Yourself in his Place 2 v. — A Terrible Temptation 2 v. — Peg Woffington 1 v. — Christie Johnstone 1 v. — A Simpleton 2 v. — The Wandering Heir 1 v. — A Woman-Hater 2 v. — Readiana 1 v. — Singleheart and Doubleface 1 v.

"Recommended to Mercy," Author of (Mrs. Houstoun).
"Recommended to Mercy" 2 v. — Zoe's "Brand" 2 v.

Reeves, Mrs.: *v.* Helen Mathers.

Rhys, Grace.
Mary Dominic 1 v. — The Wooing of Sheila 1 v.

Rice, James: *v.* Walter Besant.

Richards, Alfred Bate, † 1876.
So very Human 3 v.

Richardson, S., † 1761.
Clarissa Harlowe 4 v.

Riddell, Mrs. (F. G. Trafford).
George Geith of Fen Court 2 v. — Maxwell Drewitt 2 v. — The Race for Wealth 2 v. — Far above Rubies 2 v. — The Earl's Promise 2 v. — Mortomley's Estate 2 v.

Ridge, W Pett.
Name of Garland 1 v. — Thanks to Sanderson 1 v.

"Rita."
Souls 1 v. — The Jesters 1 v. — The Masqueraders 2 v. — Queer Lady Judas 2 v. — Prince Charming 1 v. — The Pointing Finger 1 v. — A Man of no Importance 1 v. — The Millionaire Girl, and Other Stories 1 v. — The House called Hurrish 1 v. — Calvary 2 v. — That is to say— 1 v. — "Half a Truth" 1 v.

Ritchie, Mrs. Anne Thackeray: *vide* Miss Thackeray.

Roberts, Miss: *vide* Author of "Mademoiselle Mori."

Robertson, Rev. F. W., † 1853.
Sermons 4 v.

Robins, Miss: *vide* Raimond.

Robinson, F.: *v.* "No Church."

Roosevelt, Theodore (Am.).
Outdoor Pastimes of an American Hunter (with Portrait) 1 v.

Ross, Charles H.
The Pretty Widow 1 v. — A London Romance 2 v.

Ross, Martin: *vide* Somerville.

Rossetti, Dante Gabriel, † 1882.
Poems 1 v. — Ballads and Sonnets 1 v.

"Roy Tellet."

The Outcasts 1 v. — A Draught of Lethe 1 v. — Pastor and Prelate 2 v.

Ruffini, J., † 1881.

Lavinia 2 v. — Doctor Antonio 1 v. — Lorenzo Benoni 1 v. — Vincenzo 2 v. — A Quiet Nook in the Jura 1 v. — The Paragreens on a Visit to Paris 1 v. — Carlino, and other Stories 1 v.

Ruskin, John, * 1819, † 1900.

Sesame and Lilies 1 v. — The Stones of Venice (with Illustrations) 2 v. — Unto this Last and Munera Pulveris 1 v. — The Seven Lamps of Architecture (with 14 Illustrations) 1 v. — Mornings in Florence 1 v. — St. Mark's Rest 1 v.

Russell, W. Clark.

A Sailor's Sweetheart 2 v. — The "Lady Maud" 2 v. — A Sea Queen 2 v.

Russell, George W. E.

Collections and Recollections. By One who has kept a Diary 2 v. — A Londoner's Log-Book 1 v.

Sala, George Augustus, † 1895.

The Seven Sons of Mammon 2 v.

Saunders, John.

Israel Mort, Overman 2 v. — The Shipowner's Daughter 2 v. — A Noble Wife 2 v.

Saunders, Katherine (Mrs. Cooper).

Joan Merryweather, and other Tales 1 v. — Gideon's Rock, and other Tales 1 v. — The High Mills 2 v. — Sebastian 1 v.

Savage, Richard Henry (Am.), † 1903.

My Official Wife 1 v. — The Little Lady of Lagunitas (with Portrait) 2 v. — Prince Schamyl's Wooing 1 v. — The Masked Venus 2 v. — Delilah of Harlem 2 v. — The Anarchist 2 v. — A Daughter of Judas 1 v. — In the Old Chateau 1 v. — Miss Devereux of the Mariquita 2 v. — Checked Through 2 v. — A Modern Corsair 2 v. — In the Swim 2 v. — The White Lady of Khaminavatka 2 v. — In the House of His Friends 2 v. — The Mystery of a Shipyard 2 v. — A Monte Cristo in Khaki 1 v.

Schreiner, Olive.

Trooper Peter Halket of Mashonaland 1 v. — Woman and Labour 1 v.

Scott, Sir Walter, † 1832.

Waverley (with Portrait) 1 v. — The Antiquary 1 v. — Ivanhoe 1 v. — Kenilworth 1 v. — Quentin Durward 1 v. — Old Mortality 1 v. — Guy Mannering 1 v. — Rob Roy 1 v. — The Pirate 1 v. — The Fortunes of Nigel 1 v. — The Black Dwarf; A Legend of Montrose 1 v. — The Bride of Lammermoor 1 v. — The Heart of Mid-Lothian 2 v. — The Monastery 1 v. — The Abbot 1 v. — Peveril of the Peak 2 v. — Poetical Works 2 v. — Woodstock 1 v. — The Fair Maid of Perth 1 v. — Anne of Geierstein 1 v.

Seeley, Prof. J. R., † 1895.

Life and Times of Stein (with a Portrait of Stein) 4 v. — The Expansion of England 1 v. — Goethe 1 v.

Sewell, Elizabeth, † 1906.

Amy Herbert 2 v. — Ursula 2 v. — A Glimpse of the World 2 v. — The Journal of a Home Life 2 v. — After Life 2 v. — The Experience of Life 2 v.

Shakespeare, William, † 1616.

Plays and Poems (with Portrait) *(Second Edition)* 7 v. — Doubtful Plays 1 v.

Shakespeare's Plays may also be had in 37 numbers, at ℳ 0,30. each number.

Sharp, William, † 1905: *v.* Miss Howard, Fiona Macleod and Swinburne.

Shelley, Percy Bysshe, † 1822.

A Selection from his Poems 1 v.

Sheppard, Nathan (Am.), † 1888.

Shut up in Paris 1 v.

Sheridan, R. B., † 1816.

The Dramatic Works 1 v.

Shorthouse, J. Henry.

John Inglesant 2 v. — Blanche, Lady Falaise 1 v.

Sidgwick, Mrs. Alfred.

The Lantern Bearers 1 v. — Anthea's Guest 1 v.

Slatin Pasha, Rudolf C., C.B.

Fire and Sword in the Sudan (with two Maps in Colours) 3 v.

Smedley, F. E.: *vide* Author of "Frank Fairlegh."

Smollett, Tobias, † 1771.

Roderick Random 1 v. — Humphry Clinker 1 v. — Peregrine Pickle 2 v.

Snaith, J. C.

Mrs. Fitz 1 v.

"Society in London," Author of.

Society in London. By a Foreign Resident 1 v.

Somerville, E. Œ., & M. Ross.

Naboth's Vineyard 1 v. — All on the Irish Shore 1 v. — Dan Russel the Fox 1 v.

"Spanish Brothers, the," Author of.
The Spanish Brothers 2 v.

Stanhope, Earl (Lord Mahon), † 1875.
The History of England 7 v. — Reign of Queen Anne 2 v.

Stanton, Theodore (Am.).
A Manual of American Literature 1 v.

Steel, Flora Annie.
The Hosts of the Lord 2 v. — In the Guardianship of God 1 v.

Steevens, G. W., † 1900.
From Capetown to Ladysmith 1 v.

Sterne, Laurence, † 1768.
Tristram Shandy 1 v. — A Sentimental Journey (with Portrait) 1 v.

Stevenson, Robert Louis, † 1894.
Treasure Island 1 v. — Dr. Jekyll and Mr. Hyde, and An Inland Voyage 1 v. — Kidnapped 1 v. — The Black Arrow 1 v. — The Master of Ballantrae 1 v. — The Merry Men, etc. 1 v. — Across the Plains, etc. 1 v. — Island Nights' Entertainments 1 v. — Catriona 1 v. — Weir of Hermiston 1 v. — St. Ives 2 v. — In the South Seas 2 v. — Tales and Fantasies 1 v.

"Still Waters," Author of (Mrs. Paul).
Still Waters 1 v. — Dorothy 1 v. — De Cressy 1 v. — Uncle Ralph 1 v. — Maiden Sisters 1 v. — Martha Brown 1 v. — Vanessa 1 v.

Stirling, M. C.: *vide* G. M. Craik.

Stockton, Frank R. (Am.), † 1902.
The House of Martha 1 v.

"Story of a Penitent Soul, the," Author of.
The Story of a Penitent Soul 1 v.

"Story of Elizabeth, the," Author of: *vide* Miss Thackeray.

Stowe, Mrs. Harriet Beecher (Am.), † 1896.
Uncle Tom's Cabin (with Portrait) 2 v. — A Key to Uncle Tom's Cabin 2 v. — Dred 2 v. — The Minister's Wooing 1 v. — Oldtown Folks 2 v.

"Sunbeam Stories," Author of: *vide* Mrs. Mackarness.

Swift, Jonathan (Dean Swift), † 1745.
Gulliver's Travels 1 v.

Swinburne, Algernon Charles, † 1909.
Atalanta in Calydon: and Lyrical Poems (edited, with an Introduction, by William Sharp) 1 v. — Love's Cross-Currents 1 v. — Chastelard and Mary Stuart 1 v.

Symonds, John Addington, † 1893.
Sketches in Italy 1 v. — New Italian Sketches 1 v.

Tallentyre, S. G.: *v.* H. S. Merriman.

Tasma.
Uncle Piper of Piper's Hill 2 v.

Tautphoeus, Baroness, † 1893.
Cyrilla 2 v. — The Initials 2 v. — Quits 2 v. — At Odds 2 v.

Taylor, Col. Meadows, † 1876.
Tara; a Mahratta Tale 3 v.

Templeton: *vide* Author of "Horace Templeton."

Tennyson, Alfred (Lord), † 1892.
Poetical Works 8 v. — Queen Mary 1 v. — Harold 1 v. — Becket; The Cup; The Falcon 1 v. — Locksley Hall, sixty Years after; The Promise of May; Tiresias and other Poems 1 v. — A Memoir. By His Son (with Portrait) 4 v.

Testament, the New: *vide* New.

Thackeray, William Makepeace, † 1863.
Vanity Fair 3 v. — Pendennis 3 v. — Miscellanies 8 v. — Henry Esmond 2 v. — The English Humourists of the Eighteenth Century 1 v. — The Newcomes 4 v. — The Virginians 4 v. — The Four Georges; Lovel the Widower 1 v. — The Adventures of Philip 2 v. — Denis Duval 1 v. — Roundabout Papers 2 v. — Catherine 1 v. — The Irish Sketch Book 2 v. — The Paris Sketch Book (with Portrait) 2 v.

Thackeray, Miss (Lady Ritchie).
The Story of Elizabeth 1 v. — The Village on the Cliff 1 v. — Old Kensington 2 v. — Bluebeard's Keys, and other Stories 1 v. — Five Old Friends 1 v. — Miss Angel 1 v. — Out of the World, and other Tales 1 v. — Fulham Lawn, and other Tales 1 v. — From an Island. A Story and some Essays 1 v. — Da Capo, and other Tales 1 v. — Madame de Sévigné; From a Stage Box; Miss Williamson's Divagations 1 v. — A Book of Sibyls 1 v. — Mrs. Dymond 2 v. — Chapters from some Memoirs 1 v.

Thomas a Kempis: *v.* Kempis.

Thomas, A. (Mrs. Pender Cudlip).

Denis Donne 2 v. — On Guard 2 v. — Walter Goring 2 v. — Played Out 2 v. — Called to Account 2 v. — Only Herself 2 v. — A Narrow Escape 2 v.

Thomson, James, † 1748.

Poetical Works (with Portrait) 1 v.

"Thoth," Author of.

Thoth 1 v.

Thurston, E. Temple.

The Greatest Wish in the World 1 v. — Mirage 1 v. — The City of Beautiful Nonsense 1 v.—The Garden of Resurrection 1 v.

"Tim," Author of.

Tim 1 v.

Trafford, F. G.: *v.* Mrs. Riddell.

Trevelyan, George Otto.

The Life and Letters of Lord Macaulay (with Portrait) 4 v. — Selections from the Writings of Lord Macaulay 2 v. — The American Revolution (with a Map) 2 v.

Trois-Etoiles: *vide* Grenville.

Trollope, Anthony, † 1882.

Doctor Thorne 2 v. — The Bertrams 2 v. — The Warden 1 v. — Barchester Towers 2 v. — Castle Richmond 2 v. — The West Indies 1 v. — Framley Parsonage 2 v. — North America 3 v. — Orley Farm 3 v. — Rachel Ray 2 v. — The Small House at Allington 3 v. — Can you forgive her? 3 v. — The Belton Estate 2 v. — Nina Balatka 1 v. — The Last Chronicle of Barset 3 v.—The Claverings 2 v. — Phineas Finn 3 v. — He knew he was right 3 v. — The Vicar of Bullhampton 2 v. — Sir Harry Hotspur of Humblethwaite 1 v. — Ralph the Heir 2 v. — The Golden Lion of Granpere 1 v. — Australia and New Zealand 3 v. — Lady Anna 2 v. — Harry Heathcote of Gangoil 1 v. — The Way we live now 4 v. — The Prime Minister 4 v. — The American Senator 3 v. — South Africa 2 v. — Is He Popenjoy? 3 v. — An Eye for an Eye 1 v. — John Caldigate 3 v. — Cousin Henry 1 v. — The Duke's Children 3 v. — Dr. Wortle's School 1 v. — Ayala's Angel 3 v. —The Fixed Period 1 v. — Marion Fay 2 v. — Kept in the Dark 1 v. — Frau Frohmann, and other Stories 1 v. — Alice Dugdale, and other Stories 1 v. — La Mère Bauche, and other Stories 1 v. — The Mistletoe Bough, and other Stories 1 v. — An Autobiography 1 v. — An Old Man's Love 1 v.

Trollope, T. Adolphus, † 1892.

The Garstangs of Garstang Grange 2 v. — A Siren 2 v.

Trowbridge, W. R. H.

The Letters of Her Mother to Elizabeth 1 v. — A Girl of the Multitude 1 v. — That Little Marquis of Brandenburg 1 v. — A Dazzling Reprobate 1 v.

Twain, Mark (Samuel L. Clemens) (Am.), † 1910.

The Adventures of Tom Sawyer 1 v. — The Innocents Abroad; or, The New Pilgrims' Progress 2 v. —A Tramp Abroad 2 v. — "Roughing it" 1 v. — The Innocents at Home 1 v. — The Prince and the Pauper 2 v. — The Stolen White Elephant, etc. 1 v. — Life on the Mississippi 2 v. — Sketches (with Portrait) 1 v. — Huckleberry Finn 2 v. — Selections from American Humour 1 v. — A Yankee at the Court of King Arthur 2 v. — The American Claimant 1 v. — The £ 1 000 000 Bank-Note and other new Stories 1 v. — Tom Sawyer Abroad 1 v. — Pudd'nhead Wilson 1 v. — Personal Recollections of Joan of Arc 2 v.—Tom Sawyer, Detective, and other Tales 1 v. — More Tramps Abroad 2 v. — The Man that corrupted Hadleyburg, etc. 2 v. — A Double-Barrelled Detective Story, etc. 1 v. — The $30,000 Bequest, and Other Stories 1 v. — Christian Science 1 v. — Captain Stormfield's Visit to Heaven & Is Shakespeare Dead? 1 v.

"Two Cosmos, the," Author of.

The Two Cosmos 1 v.

Vachell, Horace Annesley.

Brothers 2 v. — The Face of Clay 1 v. — Her Son 1 v. — The Hill 1 v. — The Waters of Jordan 1 v. — An Impending Sword 1 v. — The Paladin 1 v. — John Verney 1 v.

"Venus and Cupid," Author of.

Venus and Cupid 1 v.

"Vèra," Author of.

Vèra 1 v. — The Hôtel du Petit St. Jean 1 v. — Blue Roses 2 v. — Within Sound of the Sea 2 v. — The Maritime Alps and their Seaboard 2 v.—Ninette 1 v.

Victoria R. I.

Leaves from the Journal of our Life in the Highlands from 1848 to 1861 1 v. — More Leaves, etc. from 1862 to 1882 1 v.

"Virginia," Author of.

Virginia 1 v.

Vizetelly, Ernest Alfred.

With Zola in England 1 v.

Walford, L. B.
Mr. Smith 2 v. — Pauline 2 v. — Cousins 2 v. — Troublesome Daughters 2 v. — Leddy Marget 1 v.

Wallace, D. Mackenzie.
Russia 3 v.

Wallace, Lew. (Am.), † 1905.
Ben-Hur 2 v.

Warburton, Eliot, † 1852.
The Crescent and the Cross 2 v. — Darien 2 v.

Ward, Mrs. Humphry.
Robert Elsmere 3 v. — David Grieve 3 v. — Miss Bretherton 1 v. — Marcella 3 v. Bessie Costrell 1 v. — Sir George Tressady 2 v. — Helbeck of Bannisdale 2 v. — Eleanor 2 v. — Lady Rose's Daughter 2 v. — The Marriage of William Ashe 2 v. — Fenwick's Career 2 v. — Diana Mallory 2 v. — Daphne; or, "Marriage à la Mode" 1 v. — Canadian Born 1 v. — The Case of Richard Meynell 2 v.

Warner, Susan *vide*: Wetherell.

Warren, Samuel, † 1877.
Diary of a late Physician 2 v. — Ten Thousand a-Year 3 v. — Now and Then 1 v. — The Lily and the Bee 1 v.

"Waterdale Neighbours, the," Author of: *v.* Justin McCarthy.

Watts-Dunton, Theodore.
Aylwin 2 v.

Wells, H. G.
The Stolen Bacillus, etc. 1 v. — The War of the Worlds 1 v. — The Invisible Man 1 v. — The Time Machine, and The Island of Doctor Moreau 1 v. — When the Sleeper Wakes 1 v. — Tales of Space and Time 1 v. — The Plattner Story, and Others 1 v. — Love and Mr. Lewisham 1 v. — The Wheels of Chance 1 v. — Anticipations 1 v. — The First Men in the Moon 1 v. — The Sea Lady 1 v. — Mankind in the Making 2 v. — Twelve Stories and a Dream 1 v. — The Food of the Gods 1 v. — A Modern Utopia 1 v. — Kipps 2 v. — In the Days of the Comet 1 v. — The Future in America 1 v. — New Worlds for Old 1 v. — The War in the Air 1 v. — Tono-Bungay 2 v. — First and Last Things 1 v. — The New Machiavelli 2 v.

Westbury, Hugh. Acte 2 v.

Wetherell, Elizabeth (Susan Warner) (Am.), † 1885.
The wide, wide World 1 v. — Queechy 2 v. — The Hills of the Shatemuc 2 v. — Say and Seal 2 v. — The Old Helmet 2 v.

Weyman, Stanley J.
The House of the Wolf 1 v. — The Story of Francis Cludde 2 v. — A Gentleman of France 2 v. — The Man in Black 1 v. — Under the Red Robe 1 v. — My Lady Rotha 2 v. — From the Memoirs of a Minister of France 1 v. — The Red Cockade 2 v. — Shrewsbury 2 v. — The Castle Inn 2 v. — Sophia 2 v. — Count Hannibal 2 v. — In Kings' Byways 1 v. — The Long Night 2 v. — The Abbess of Vlaye 2 v. — Starvecrow Farm 2 v. — Chippinge 2 v. — Laid up in Lavender 1 v.

Wharton, Edith (Am.).
The House of Mirth 2 v. — The Fruit of the Tree 2 v.

"Whim, a," Author of.
A Whim, and its Consequences 1 v.

Whitby, Beatrice.
The Awakening of Mary Fenwick 2 v. — In the Suntime of her Youth 2 v.

White, Percy.
Mr. Bailey-Martin 1 v. — The West End 2 v. — The New Christians 1 v. — Park Lane 2 v. — The Countess and The King's Diary 1 v. — The Triumph of Mrs. St. George 2 v. — A Millionaire's Daughter 1 v. — A Passionate Pilgrim 1 v. — The System 2 v. — The Patient Man 1 v. — Mr. John Strood 1 v. — The Eight Guests 2 v. — Mr. Strudge 1 v. — Love and the Poor Suitor 1 v. — The House of Intrigue 1 v. — Love and the Wise Men 1 v. — An Averted Marriage 1 v. — The Lost Halo 1 v. — The Broken Phial 1 v.

White, Walter.
Holidays in Tyrol 1 v.

Whiteing, Richard.
The Island; or, An Adventure of a Person of Quality 1 v. — No. 5 John Street 1 v. — The Life of Paris 1 v. — The Yellow Van 1 v. — Ring in the New 1 v. — All Moonshine 1 v. — Little People 1 v.

Whitman, Sidney.
Imperial Germany 1 v. — The Realm of the Habsburgs 1 v. — Teuton Studies 1 v. — Reminiscences of the King of Roumania 1 v. — Conversations with Prince Bismarck 1 v. — Life of the Emperor Frederick 2 v.

"Who Breaks—Pays," Author of: *vide* Mrs. Jenkin.

Whyte Melville, George J.: *vide* Melville.

Wiggin, Kate Douglas (Am.).
Timothy's Quest 1 v. — A Cathedral Courtship, and Penelope's English Ex-

periences 1 v. — Penelope's Irish Experiences 1 v. — Rebecca of Sunnybrook Farm 1 v. — Rose o' the River 1 v. — New Chronicles of Rebecca 1 v. — The Old Peabody Pew, and Susanna and Sue 1 v. — Mother Carey 1 v.

K. D. Wiggin, M. & J. Findlater, & Allan McAulay.

The Affair at the Inn 1 v. — Robinetta 1 v.

Wilde, Oscar, † 1900.

The Picture of Dorian Gray 1 v. — De Profundis and The Ballad of Reading Gaol 1 v. — A House of Pomegranates 1 v. — Lord Arthur Savile's Crime, and Other Prose Pieces 1 v.—Lady Windermere's Fan 1 v.— An Ideal Husband 1 v.— Salome 1 v. — The Happy Prince, and Other Tales 1 v. — A Woman of No Importance 1 v. — The Importance of Being Earnest 1 v. — Poems 1 v.

Wilkins, Mary E. (Am.).

Pembroke 1 v. — Madelon 1 v. — Jerome 2 v. — Silence, and other Stories 1 v. — The Love of Parson Lord, etc. 1 v.

Williamson, C. N. & A. M. (Am.).

The Lightning Conductor 1 v.—Lady Betty across the Water 1 v.—The Motor Maid 1 v. — Lord Loveland discovers America 1 v. — The Golden Silence 2 v.

Wills, C. J., *vide* F. C. Philips.

Winter, Mrs. J. S.

Regimental Legends 1 v.

Wood, C.: *vide* "Buried Alone."

Wood, H. F.

The Passenger from Scotland Yard 1 v.

Wood, Mrs. Henry (Johnny Ludlow), † 1887.

East Lynne 3 v. — The Channings 2 v. — Mrs. Halliburton's Troubles 2 v. — Verner's Pride 3 v.—The Shadow of Ashlydyat 3 v. — Trevlyn Hold 2 v. — Lord Oakburn's Daughters 2 v. — Oswald Cray 2 v. — Mildred Arkell 2 v. — St. Martin's Eve 2 v. — Elster's Folly 2 v. — Lady Adelaide's Oath 2 v. — Orville College 1 v. — A Life's Secret 1 v. — The Red Court Farm 2 v. — Anne Hereford 2 v. — Roland Yorke 2 v. — George Canterbury's Will 2 v. — Bessy Rane 2 v. — Dene Hollow 2 v. — The Foggy Night at Offord; Martyn Ware's Temptation; The Night-Walk over the Mill Stream 1 v. — Within the Maze 2 v. — The Master of Greylands 2 v. — Johnny Ludlow 2 v. — Told in the Twilight 2 v. — Adam Grainger 1 v. — Edina 2 v. — Pomeroy Abbey 2 v. — Court Netherleigh 2 v. — (The following by Johnny Ludlow): Lost in the Post, and Other Tales 1 v.—A Tale of Sin, and Other Tales 1 v. — Anne, and Other Tales 1 v. — The Mystery of Jessy Page, etc. 1 v. — Helen Whitney's Wedding, etc. 1 v.—The Story of Dorothy Grape, etc. 1 v.

Woodroffe, Daniel.

Tangled Trinities 1 v.–The Beauty-Shop 1 v.

Woods, Margaret L.

A Village Tragedy 1 v. — The Vagabonds 1 v. — Sons of the Sword 2 v. — The Invader 1 v.

Wordsworth, William, † 1850.

Select Poetical Works 2 v.

Wraxall, Lascelles, † 1865.

Wild Oats 1 v.

Yates, Edmund, † 1894.

Land at Last 2 v. — Broken to Harness 2 v. — The Forlorn Hope 2 v. — Black Sheep 2 v. — The Rock Ahead 2 v. — Wrecked in Port 2 v. — Dr. Wainwright's Patient 2 v. — Nobody's Fortune 2 v. — Castaway 2 v. — A Waiting Race 2 v. — The yellow Flag 2 v. — The Impending Sword 2 v. — Two, by Tricks 1 v. — A Silent Witness 2 v. — Recollections and Experiences 2 v.

Yeats: *vide* Levett-Yeats.

Yonge, Charlotte M., † 1901.

The Heir of Redclyffe 2 v. — Heartsease 2 v. — The Daisy Chain 2 v. — Dynevor Terrace 2 v. — Hopes and Fears 2 v. — The Young Step-Mother 2 v. — The Trial 2 v. — The Clever Woman of the Family 2 v. — The Dove in the Eagle's Nest 2 v. — The Danvers Papers; The Prince and the Page 1 v. — The Chaplet of Pearls 2 v. — The two Guardians 1 v. — The Caged Lion 2 v. — The Pillars of the House 5 v. — Lady Hester 1 v. — My Young Alcides 2 v. — The Three Brides 2 v. — Womankind 2 v. — Magnum Bonum 2 v. — Love and Life 1 v. — Unknown to History 2 v. — Stray Pearls (with Portrait) 2 v. — The Armourer's Prentices 2 v. — The Two Sides of the Shield 2 v. — Nuttie's Father 2 v. — Beechcroft at Rockstone 2 v. — A Reputed Changeling 2 v. — Two Penniless Princesses 1 v. — That Stick 1 v. — Grisly Grisell 1 v. — The Long Vacation 2 v. — Modern Broods 1 v.

"Young Mistley," Author of: *vide* Henry Seton Merriman.

Zangwill, I.

Dreamers of the Ghetto 2 v. — Ghetto Comedies 2 v. — Ghetto Tragedies 2 v.

"*Z. Z.*"

The World and a Man 2 v.

Series for the Young.

30 Volumes. Published with Continental Copyright on the same conditions as the Collection of English and American Authors. Vide p. 1.

— Price 1 M. 60 Pf. or 2 Fr. per Volume. —

Barker, Lady (Lady Broome).
Stories About:— 1 v.

Charlesworth, Maria Louisa, † 1880.
Ministering Children 1 v.

Craik, Mrs. (Miss Mulock), † 1887.
Our Year 1 v. — Three Tales for Boys 1 v. — Three Tales for Girls 1 v.

Craik, Georgiana M. (Mrs. May).
Cousin Trix, and her Welcome Tales 1 v.

Edgeworth, Maria, † 1849.
Moral Tales 1 v. — Popular Tales 2 v.

Kavanagh, Bridget & Julia, † 1877.
The Pearl Fountain, and other Fairy-Tales 1 v.

Lamb, Charles & Mary, † 1834 and 1847.
Tales from Shakspeare 1 v.

Marryat, Captain, † 1848.
Masterman Ready 1 v.

Marshall, Mrs. Emma, † 1899.
Rex and Regina 1 v.

Montgomery, Florence.
The Town-Crier; to which is added: The Children with the Indian-Rubber Ball 1 v.

"Ruth and her Friends," Author of.
Ruth and her Friends. A Story for Girls 1 v.

Wood, Mrs. Henry, † 1887.
William Allair 1 v.

Yonge, Charlotte M., † 1901.
Kenneth; or, the Rear-Guard of the Grand Army 1 v. — The Little Duke. Ben Sylvester's Word 1 v. — The Stokesley Secret 1 v. — Countess Kate 1 v. — A Book of Golden Deeds 2 v. — Friarswood Post-Office 1 v. — Henrietta's Wish 1 v. — Kings of England 1 v. — The Lances of Lynwood; the Pigeon Pie 1 v. —P's and Q's 1 v.—Aunt Charlotte's Stories of English History 1 v. — Bye-Words 1 v.— Lads and Lasses of Langley, etc. 1 v.

Collection of German Authors.

51 Volumes. Translations from the German, published with universal copyright. These volumes may be imported into any country.

— Price 1 M. 60 Pf. or 2 Fr. per Volume. —

Auerbach, Berthold, † 1882.
On the Heights, *(Second Edition)* 3 v. — Brigitta 1 v. — Spinoza 2 v.

Ebers, Georg, † 1898.
An Egyptian Princess 2 v. — Uarda 2 v. — Homo Sum 2 v. — The Sisters [Die Schwestern] 2 v. — Joshua 2 v. — Per Aspera 2 v.

Fouqué, De la Motte, † 1843.
Undine, Sintram, etc. 1 v.

Freiligrath, Ferdinand, † 1876.
Poems *(Second Edition)* 1 v.

Görlach, Wilhelm.
Prince Bismarck (with Portrait) 1 v.

Goethe, W. v., † 1832.
Faust 1 v. — Wilhelm Meister's Apprenticeship 2 v.

Gutzkow, Karl, † 1878.
Through Night to Light 1 v.

Hackländer, F. W., † 1877.
Behind the Counter [Handel und Wandel] 1 v.

Hauff, Wilhelm, † 1827.
Three Tales 1 v.

Heyse, Paul.
L'Arrabiata, etc. 1 v. — The Dead Lake, etc. 1 v. — Barbarossa, etc. 1 v.

Hillern, Wilhelmine von.
The Vulture Maiden [die Geier-Wally] 1 v. — The Hour will come 2 v.

Kohn, Salomon.
Gabriel 1 v.

Lessing, G. E., † 1781.
Nathan the Wise and Emilia Galotti 1 v.

Lewald, Fanny, † 1889.
Stella 2 v.

Marlitt, E., † 1887.
The Princess of the Moor [das Haideprinzesschen] 2 v.

Nathusius, Maria, † 1857.
Joachim v. Kamern, and Diary of a Poor Young Lady 1 v.

Reuter, Fritz, † 1874.
In the Year '13 1 v. — An old Story of my Farming Days [Ut mine Stromtid] 3 v.

Richter, J. P. Friedrich (Jean Paul), † 1825.
Flower, Fruit and Thorn Pieces 2 v.

Scheffel, Victor von, † 1886.
Ekkehard 2 v.

Taylor, George.
Klytia 2 v.

Zschokke, Heinrich, † 1848.
The Princess of Brunswick-Wolfenbüttel, etc. 1 v.

Students' Series for School, College, and Home.

Ausgaben
mit deutschen Anmerkungen und Special-Wörterbüchern.

Br. = Broschiert. Kart. = Kartoniert.

Bulwer, Edward, Lord Lytton, † 1873.
The Lady of Lyons. Von Dr. *Fritz Bischoff.* Br. ℳ 0,50. Kart. ℳ 0,60.

Burnett, Frances Hodgson (Am.).
Little Lord Fauntleroy. Von Dr. *Ernst Groth.* Br. ℳ 1,50. Kart. ℳ 1,60. — Anmerkungen und Wörterbuch. Br. ℳ 0,40.
Sara Crewe. Von *Bertha Connell.* Br. ℳ 0,50. Kart. ℳ 0,60. — Anmerkungen und Wörterbuch. Br. ℳ 0,40.

Carlyle, Thomas, † 1881.
The Reign of Terror (French Revolution). Von Dr. *Ludwig Herrig.* Br. ℳ 1,00. Kart. ℳ 1,10.

Craik, Mrs. (Miss Mulock), † 1887.
A Hero. A Tale for Boys. Von Dr. *Otto Dost.* Br. ℳ 0,80. Kart. ℳ 0,90.— Wörterbuch. Br. ℳ 0,40.

Dickens, Charles, † 1870.
Sketches. First Series. Von Dr. *A. Hoppe.* Br. ℳ 1,20. Kart. ℳ 1,30.
Sketches. Second Series. Von Dr. *A. Hoppe.* Br. ℳ 1,40. Kart. ℳ 1,50.—Wörterbuch (First and Second Series). Br. ℳ 1,00.
A Christmas Carol in Prose. Being a Ghost Story of Christmas. Von Dr. *G. Tanger.* Br. ℳ 1,00. Kart. ℳ 1,10.

Eliot, George (Miss Evans—Mrs. Cross), † 1880.
The Mill on the Floss. Von Dr. *H. Conrad.* Br. ℳ 1,70. Kart. ℳ 1,80.

Ewing, Juliana Horatia, † 1885.
Jackanapes. Von *E. Roos.* Br. ℳ 0,50. Kart. ℳ 0,60. —Wörterbuch. Br. ℳ 0,20.
The Brownies; and The Land of Lost Toys. Von Dr. *A. Müller.* Br. ℳ 0,60. Kart. ℳ 0,70.—Wörterbuch Br. ℳ 0,30.
Timothy's Shoes; An Idyll of the Wood; Benjy in Beastland. Von *E. Roos.* Br. ℳ 0,70. Kart. ℳ 0,80. — Wörterbuch. Br. ℳ 0,30.

Franklin, Benjamin (Am.), † 1790.
His Autobiography. Von Dr. *Karl Feyerabend.* I. Teil. Die Jugendjahre (1706—1730). Br. ℳ 1,00. Kart. ℳ 1,10
II. Teil. Die Mannesjahre (1731 bis 1757). Mit einer Beigabe: The Way to Wealth. Von Dr. *Karl Feyerabend.* Br. ℳ 1,20. Kart. ℳ 1,30.

Freeman, Edward A. † 1892.
Three Historical Essays. Von Dr. *C. Balzer.* Br. ℳ 0,70. Kart. ℳ 0,80.

Harte, Bret (Am.), † 1902.
Tales of the Argonauts. Von Dr. *G. Tanger.* Br. ℳ 1,40. Kart. ℳ 1,50.

Hawthorne, Nathaniel (Am.), † 1864.

Wonder Book for Boys and Girls. Von *E. Roos.* Br. ℳ 0,70. Kart. ℳ 0,80. — Anmerkungen und Wörterbuch. Br. ℳ 0,40.

Hughes, Thomas, † 1898.

Tom Brown's School Days. Von Dr. *I. Schmidt.* 2 Parts. Br. ℳ 3,00. Kart. ℳ 3,20. Part I. apart. Br. ℳ 1,70. Kart. ℳ 1,80. Part. II. apart. Br. ℳ 1,30. Kart. ℳ 1,40.

Longfellow, Henry Wadsworth (Am.), † 1882.

Tales of a Wayside Inn. Von Dr. *H. Varnhagen.* 2 Bände. Br. ℳ 2,00. Kart. ℳ 2,20. 1. Band apart. Br. ℳ 1,00. Kart. ℳ 1,10. 2. Band apart. Br. ℳ 1,00. Kart. ℳ 1,10.

Macaulay, Lord, Thomas Babington, † 1859.

England before the Restoration. (History of England. Chapter I.) Von Dr. *W. Ihne.* Br. ℳ 0,70. Kart. ℳ 0,80.

England under Charles the Second. (History of England. Chapter II.) Von Dr. *W. Ihne.* Br. ℳ 1,00. Kart. ℳ 1,10.

The Rebellions of Argyle and Monmouth. (History of England. Chapter V.) Von Dr. *Immanuel Schmidt.* Br. ℳ 1,00. Kart. ℳ 1,10.

Lord Clive. (Histor. Essay.) Von Prof. Dr. *R. Thum.* Br. ℳ 1,40. Kart. ℳ 1,50.

Ranke's History of the Popes. (Historical Essay.) Von Prof. Dr. *R. Thum.* Br. ℳ 0,60. Kart. ℳ 0,70.

Warren Hastings. (Historical Essay.) Von Prof. Dr. *R. Thum.* Br. ℳ 1,50. Kart. ℳ 1,60.

McCarthy, Justin.

The Indian Mutiny. (Chap. 32—35 of "A History of our own Times.") Von Dr. *A. Hamann.* Br. ℳ 0,60. Kart. ℳ 0,70. — Wörterbuch. Br. ℳ 0,20.

Montgomery, Florence.

Misunderstood. Von Dr. *R. Palm.* Br. ℳ 1,60. Kart. ℳ 1,70. — Wörterbuch. Br. ℳ 0,40.

Scott, Sir Walter, † 1832.

The Talisman. Von Dr. *R. Dressel.* Br. ℳ 1,60. Kart. ℳ 1,70.

Tales of a Grandfather. First Series. Von Dr. *H. Löschhorn.* Br. ℳ 1,50. Kart. ℳ 1,60. — Wörterbuch. Br. ℳ 0,50.

Tales of a Grandfather. Second Series. Von Dr. *H. Löschhorn.* Br. ℳ 1,70. Kart. ℳ 1,80.

Shakespeare, William, † 1616.

Twelfth Night; or, What you will. Von Dr. *H. Conrad.* Br. ℳ 1,40. Kart. ℳ 1,50.

Julius Cæsar. Von Dr. *Immanuel Schmidt.* Br. ℳ 1,00. Kart. ℳ 1,10.

Macbeth. Von Dr. *Immanuel Schmidt.* Br. ℳ 1,00. Kart. ℳ 1,10.

Stanhope, Earl (Lord Mahon), † 1875.

Prince Charles Stuart. (History of England from the Peace of Utrecht to the Peace of Versailles. 1713—1783.) Von Dr. *Martin Krummacher.* Br. ℳ 1,20. Kart. ℳ 1,30.

The Seven Years' War. Von Dr. *M. Krummacher.* Br. ℳ 1,20. Kart. ℳ 1,30.

Tennyson, Alfred Lord, † 1892.

Enoch Arden and other Poems. Von Dr. *A. Hamann.* Br. ℳ 0,70. Kart. ℳ 0,80. — Wörterbuch. Br. ℳ 0,20.

Thackeray, W. M. † 1863.

Samuel Titmarsh and The great Hoggarty Diamond. Von *George Boyle.* Br. ℳ 1,20. Kart. ℳ 1,30.

Yonge, Charlotte M., † 1901.

The Little Duke, or, Richard the Fearless. Von *E. Roos.* Br. ℳ 0,90. Kart. ℳ 1,00. — Wörterbuch. Br. ℳ 0,20.

Manuals of Conversation (same size as Tauchnitz Edition).

Each Volume, bound ℳ 2,25.

Für Deutsche.

Englische Conversationssprache von *A. Schlessing.* 4. Stereotypaufl.

Französische Conversationssprache von *L. Rollin.* 2. Stereotypaufl.

Russische Conversationssprache von Dr. *Z. Koiransky.*

For English students.

German Language of Conversation by *A. Schlessing.*

À l'usage des étudiants français.

Conversation Allemande par MM. *L. Rollin* et *Wolfgang Weber.*

Made in the USA
Monee, IL
05 November 2026

67961489R00203